Cambridge Elements

Elements in the Foundations of Contemporary Physics
edited by
Richard Dawid
Stockholm University
James Wells
University of Michigan, Ann Arbor

PROBING THE CONSISTENCY OF QUANTUM FIELD THEORY I

From Nonconvergence to Haag's Theorem (1949–1954)

Alexander S. Blum
Munich Center for Mathematical Philosophy and Max Planck Institute for the History of Science

Shaftesbury Road, Cambridge CB2 8EA, United Kingdom

One Liberty Plaza, 20th Floor, New York, NY 10006, USA

477 Williamstown Road, Port Melbourne, VIC 3207, Australia

314–321, 3rd Floor, Plot 3, Splendor Forum, Jasola District Centre, New Delhi – 110025, India

103 Penang Road, #05–06/07, Visioncrest Commercial, Singapore 238467

Cambridge University Press is part of Cambridge University Press & Assessment, a department of the University of Cambridge.

We share the University's mission to contribute to society through the pursuit of education, learning and research at the highest international levels of excellence.

www.cambridge.org
Information on this title: www.cambridge.org/9781009619271

DOI: 10.1017/9781009265362

© Alexander S. Blum 2026

This publication is in copyright. Subject to statutory exception and to the provisions of relevant collective licensing agreements, with the exception of the Creative Commons version the link for which is provided below, no reproduction of any part may take place without the written permission of Cambridge University Press & Assessment.

An online version of this work is published at doi.org/10.1017/9781009265362 under a Creative Commons Open Access license CC-BY-ND 4.0 which permits re-use, distribution and reproduction in any medium providing appropriate credit to the original work is given. You may not distribute derivative works without permission. To view a copy of this license, visit https://creativecommons.org/licenses/by-nd/4.0

When citing this work, please include a reference to the DOI 10.1017/9781009265362

First published 2026

A catalogue record for this publication is available from the British Library

ISBN 978-1-009-61927-1 Hardback
ISBN 978-1-009-26533-1 Paperback
ISSN 2752-3039 (online)
ISSN 2752-3020 (print)

Cambridge University Press & Assessment has no responsibility for the persistence or accuracy of URLs for external or third-party internet websites referred to in this publication and does not guarantee that any content on such websites is, or will remain, accurate or appropriate.

For EU product safety concerns, contact us at Calle de José Abascal, 56, 1°, 28003 Madrid, Spain, or email eugpsr@cambridge.org

Probing the Consistency of Quantum Field Theory I

From Nonconvergence to Haag's Theorem (1949–1954)

Elements in the Foundations of Contemporary Physics

DOI: 10.1017/9781009265362
First published online: January 2026

Alexander S. Blum
Munich Center for Mathematical Philosophy and Max Planck Institute for the History of Science

Author for correspondence: Alexander S. Blum,
ablum@mpiwg-berlin.mpg.de
alexander.blum@lmu.de

Abstract: This two-volume Element reconstructs and analyzes the historical debates on whether renormalized quantum field theory is a mathematically consistent theory. This volume covers the years immediately following the development of renormalized quantum electrodynamics. It begins with the realization that perturbation theory cannot serve as the foundation for a proof of consistency, due to the nonconvergence of the perturbation series. Various attempts at a nonperturbative formulation of quantum field theory are discussed, including the Schwinger–Dyson equations, Gunnar Källén's nonperturbative renormalization, the renormalization group of Murray Gell-Mann and Francis Low, and, in the last section, early axiomatic quantum field theory. The volume concludes with the establishment of Haag's theorem, which proved that even the Hilbert space of perturbation theory is an inadequate foundation for a consistent theory. This title is also available as Open Access on Cambridge Core.

Keywords: quantum field theory, history of physics, inconsistency, renormalization, philosophy of physics

© Alexander S. Blum 2026

ISBNs: 9781009619271 (HB), 9781009265331 (PB), 9781009265362 (OC)
ISSNs: 2752-3039 (online), 2752-3020 (print)

Contents

1	Introduction	1
2	The Divergence of the Perturbation Series	9
3	The Search for Nonperturbative Solutions	37
4	Infinite Renormalization and UV Behavior	55
5	The Axiomatic Approach	68
	References	96

1 Introduction

Will the final theory of physics be written in the language of quantum field theory (QFT)? This was the question debated by two men at the Institute for Advanced Study (IAS) in Princeton on January 22, 1949. The two interlocutors could hardly have been more aptly chosen: In 1930, the older of the two, J. Robert Oppenheimer – now director of the institute and known around the world as the father of the atomic bomb, but back then a young postdoc in Zurich – had established that QFT was fatally flawed. He had calculated the corrected values for the spectral lines of hydrogen in the quantum field theory of electrodynamics and had found them to be infinite (Oppenheimer, 1930). Some 20 years later, Freeman J. Dyson – a PhD student of sorts, though he never ended up graduating – had just written the definitive paper on how these infinities could be removed in a systematic procedure called renormalization, delivering a theory of quantum electrodynamics (QED) that returned sensible and empirically highly accurate predictions, in particular for the spectral lines of hydrogen (Dyson, 1949a). One man had led fundamental physics into two decades of soul-searching; the other had led it out again.

Neither of the two was particularly happy with renormalized quantum field theory. In a letter to his parents – our only source on the discussion with Oppenheimer (Dyson, 2018, p. 136) – Dyson remarked that "we are agreed that the existing methods of field theory are not satisfactory and must ultimately be scrapped in favour of a theory which is physically more intelligible and less arbitrary. We also agree that the final theory should explain why there are the various types of particles we see and no more." However, Oppenheimer and Dyson disagreed on how and when this "final theory" would be achieved. According to Dyson, "Oppenheimer believes that the nature of the nuclear forces will itself give us enough information on which to build the new theory," that is, he believed that the next great transformation of fundamental physics (and presumably the final one) was right around the corner and was to be found in the detailed study of nuclear physics, a scientific endeavor that the postwar United States was pursuing with unparalleled resources. Dyson, on the other hand, believed that "we shall be able to give a complete account of the nucleus on the basis of the present field theory," that is, that a renormalized QFT for the nuclear interactions could be constructed following the example of QED and that "the discovery of a finally satisfactory theory of elementary particles will be a much deeper problem than those we are tackling at present and may very well not be achieved within the framework of microscopic physics."

It is uncontroversial that history proved Oppenheimer wrong: less than 30 years later, the Standard Model of Particle Physics was established, encompassing quantum field theories for both the strong and weak nuclear interactions. It is, however, hard to see the events that followed the discussion in Princeton as a vindication of Dyson's position. The program that Dyson pursued afterwards was equally a failure: he attempted to show that renormalized QED was a "complete and consistent theory" (Dyson, 1952, p. 631) that could thus serve as a blueprint for quantum field theories of the nuclear interactions – only to find the consistency of QED to be an elusive statement that he could neither prove nor disprove. Thus began one of the most curious stories in the history of modern physics. Quantum electrodynamics came to be hailed as the most precise theory in the history of physics, famously delivering a theoretical prediction for the g-factor of the anomalous magnetic moment of the electron that has been confirmed experimentally to 5 parts in a trillion (Hanneke et al., 2011). Richard Feynman considered QED (which, of course, he himself had helped develop) as "the jewel of physics – our proudest possession" (Feynman, 1985, p. 8). Yet Dyson and those that followed him were not just unable to prove the mathematical consistency of this theory, they were equally unable to prove its inconsistency. This two-volume Element recounts the story of these attempts, of how our arguably best physical theory ended up, and remained, in mathematical limbo. It follows this story from Dyson's first forays in the late 1940s to Kenneth Wilson's renormalization group analysis in the mid 1960s.

Over the course of these two decades, we observe a considerable amount of variety and change: different notions of consistency, different methods for probing it, different conclusions regarding the consistency of QED. It is thus far from obvious what should be included in a history of – as promised in this Element's title – the attempts at probing the consistency of QED. Given the fact that this is largely uncharted historiographical territory, given the technical complexity of the issues, and given the exponential growth of post-WWII high-energy physics, a certain amount of arbitrariness, idiosyncrasy, and teleology in the selection of material is likely inevitable. Still, I hope that there are clearly discernible themes running through the narrative presented here, and I will attempt to sketch them in the following.

It is easiest to begin by stating what this Element is not about. As we have seen, both Oppenheimer and Dyson were "agreed" that renormalized QED was unsatisfactory in some profound sense. This discontent was primarily due to the renormalization procedure itself and was shared to some extent by most physicists at the time. Many physicists continued to harbor such doubts for a long time, most prominently Paul Dirac, one of the founding fathers of QED. Indeed, Dirac's last paper (published posthumously) was entitled

The Inadequacies of Quantum Field Theory. But it was merely concerned with criticizing the renormalization procedure as "these rules for subtracting infinities" or "just a set of working rules, and not a complete dynamical theory at all" (Dirac, 1987, p. 196). This form of criticism is widely considered to have been made obsolete by the Wilsonian interpretation of renormalization and the rise of the effective field theory paradigm, which took shape in the 1970s (Cao and Schweber, 1993). In any case, it is not the sort of criticism I will be discussing in this Element. Dyson, for one, acknowledged the imperfections of the renormalization procedure, but still set out to prove the consistency of the renormalized theory. The inadequacies of QFT that I will address here, such as the nonconvergence of the perturbation series discovered by Dyson, are not inadequacies of the renormalization procedure, but rather inadequacies of QED that persisted even if one took the renormalization procedure for granted. To be clear, many of the physicists discussed in this Element explicitly attempted to prove QED *inconsistent* and were motivated to do so by their dissatisfaction with the renormalization procedure. But these physicists – such as, most prominently, Wolfgang Pauli – attempted to undermine renormalization by proving the inconsistency of renormalized QED, not, like Dirac, by merely pooh-poohing the renormalization procedure.

Our focus will thus be on physicists who took the existing structure of QED for granted and *analyzed* this structure to ascertain *whether* it conformed to certain criteria of consistency. This excludes the attempts at abandoning the field-theoretic approach to microscopic physics altogether, such as, most notably, the autonomous S-Matrix program (Cushing, 1990). It also excludes the many attempts at *modifying* the structure of QED in order to *make sure that* it conforms to certain criteria of consistency. This line between analyzing existing structures and constructing new ones is not always easy to draw. We will see the emergence of new approximation methods and renormalization procedures; we will also see the construction of new toy models, explicitly designed to reproduce certain structural features of QED. We will even see early attempts at constructing an entirely new, manifestly consistent (or inconsistent), axiomatic foundation for QFT – an approach that grew into an entire subdiscipline of its own. This Element aims to showcase the impact these innovations had on the consistency question, while minimizing discussion of their wider relevance.

There is one thing that all of these varied approaches to probing the consistency of QFT had in common: they were an objective failure. No hard-and-fast inconsistency proof was obtained, nor was the consistency of QFT conclusively proven. In fact, you can win yourself one million dollars from the Clay Mathematics Institute by "producing a *mathematically complete* example of quantum gauge field theory in four-dimensional space-time."[1] What, then, can we learn from this history of failure?

To answer this question, we first need to consider what sorts of insights physicists had expected to gain from investigating the consistency of QED – had they been successful in proving either the consistency or the inconsistency of the theory. We will, of course, encounter some individual and very idiosyncratic expectations and motivations throughout this Element; but some more general observations can be made. At first glance, one might well consider either outcome of these investigations as trivial. What could a proof of mathematical consistency add to the theory's flawless empirical track record? Conversely, how could a proof of mathematical inconsistency undermine a theory that was anyhow awaiting repeal from a theory of the nuclear forces? The modern reader may even find the entire preoccupation with inconsistency misplaced, having come to accept inconsistency in physical theories as a fact of life.[2] But we must not fall into the trap of misreading this whole debate as a purely philosophical one. As Dyson's letter already suggests, the consistency of QED itself was not the fundamental issue; the central problem was always the construction of a theory of the strong nuclear interactions.[3]

What was the role of the (in)consistency of QED in deciding this issue? The answer is rather simple: a consistency proof would have established QED as a blueprint for a theory of the strong nuclear interactions. Now, of course, QED *had* been used as a blueprint for theories of the nuclear interactions early on; both Fermi's theory of beta decay and Yukawa's meson theory had been crafted in explicit analogy to QED.[4] Fermi's theory had long been known to exhibit perturbative divergences that were worse than those of QED (Blum, 2017, section 1.1), and it was soon found that the methods used to renormalize QED could not be applied to it (Kamefuchi, 1951). To a zeroth approximation, the

---

[1] This (without the emphasis I added) is to be found on page 5 of the official problem description by Arthur Jaffe and Edward Witten. www.claymath.org/sites/default/files/yangmills.pdf.

[2] Examples of this view can be found in Meheus (2002), where, e.g., Thomas Nickles (2002, p. 2) states that "inconsistency problems... are less serious than everyone used to think, precisely because they are now anticipated products of ongoing, self-corrective investigations and neither productive of general intellectual disaster nor necessarily indicative of personal or methodological failure."

[3] The reader may also wonder about the problem of constructing a quantum field theory of gravity. In 1950, this was considered both far more remote and far less problematic than it is today, and it will appear only once or twice as a sidenote in this Element (Blum and Rickles, 2018; Rickles, 2020).

[4] Fermi (1934, p. 161) explicitly characterized his theory as "a theory of the emission of light particles from the nucleus in analogy to the theory of the emission of a light quantum." Similarly, Yukawa (1935, p. 48) characterized his theory as describing the interaction between nucleons "by means of a field of force just as the interaction between the charged particles is described by the electromagnetic field."

history of the weak interactions can be read as the search for a better, renormalizable (and thus more QED-like) model, culminating in the construction of the electroweak Standard Model. That development is rather orthogonal to the focus of this Element, and the weak interaction will play little role in what follows.[5]

The problem with the strong interaction was quite different. There were versions of Yukawa's theory that were clearly renormalizable. The main difference between the strong and the electromagnetic interactions was that the former are – and this was evident long before their current name was adopted – very strong in spite of their short range, that there is an "extraordinary tendency of... nuclei to react with each other as soon as direct contact is established" (Bohr, 1936, p. 344). But the calculations that underlay (and to this day underlie) the successful empirical predictions of QED relied on a *weak* coupling approximation. In fact, the entire renormalization procedure, which removed the infinities appearing in the theory, was formulated only within the framework of this perturbative approximation. One could, of course, simply ignore this difficulty and apply QED-style perturbation theory to Yukawa's field theory of the strong interaction, but this approach did not meet with much success.[6] Probing the consistency of QED thus meant finding out whether there was actually a full-fledged theory behind this approximation scheme – something that was being approximated, a framework that could be extended beyond the case of weak coupling to describe the strong interaction and ultimately provide the language for a final theory.

Such a framework was not found. Instead, physicists discovered a cornucopia of problems with renormalized QED, including the divergence of the perturbation series, Haag's theorem, and the Landau pole. These issues continue to be central in discussions on QFT today. By introducing these problematic aspects of QFT not in an abstract way, but by showing how they actually arose *in practice*, this Element also hopes to provide some clarification regarding the foundational status of QFT – not by asserting what the *real* problems of

---

[5] The distinction between the strong and the weak nuclear interactions was of course only gradually emerging during the period under study. We find it fully codified using modern terminology in Gell-Mann (1956). The sharp distinction arose from the need to categorize interactions according to which quantities (isospin, strangeness) they conserve. I would like to thank Arianna Borrelli for a helpful discussion of this point.

[6] To pick an example more or less at random, consider a paper by Sidney Borowitz and Walter Kohn, which concludes by stating: "Some quantitative aspects of the theory are correct but the quantitative agreement with experiment is very poor. The large values of the coupling constant required to fit the experimental data show that the results obtained by our second order perturbation calculation are qualitatively unreliable and indicate that higher order calculations would not be promising. To test the correctness of the meson hypothesis, some method more suitable for dealing with strong coupling is required" (Borowitz and Kohn, 1949, p. 827).

QFT are, but by emphasizing what was (and is) a problem for whom trying to do what. Once readers have seen these problems in action, they will hopefully be able to form a clearer opinion of their respective significance.

None of the problems discussed in this Element amounted to a full-fledged inconsistency proof (though several claims to that effect were made), but they contributed to a widespread distrust of QFT. This distrust led to the exploration – and even the dominance – of non–field-theoretic approaches to the strong interaction beginning in the mid 1950s, such as the aforementioned S-Matrix approach. However, in probing the consistency of QED, physicists gained more than just a typology of problems. Four key insights will emerge in the course of this Element: (i) the consistency of QED was intimately connected to its short-distance behavior; (ii) at short distances, it was charge renormalization (rather than mass renormalization, i.e., self-energy) that played the critical role in assessing the consistency of QED; (iii) there were several shades of gray between the black and white of consistency and inconsistency; and (iv) on this spectrum of consistency, there was room for models that fared far better than QED, in contrast to what both Oppenheimer and Dyson had expected.

It was hardly unreasonable to expect that QED would be the optimal field theory; not only did it originate from Maxwell's electrodynamics, the *original* field theory, but it had also just celebrated the splendid triumphs of the late 1940s, which earned Richard Feynman, Julian Schwinger, and Sin-Itiro Tomonaga the Nobel Prize in 1965. This triumph was vividly depicted by Sam Schweber (1994), and this Element can be considered a sequel of sorts to his *QED*, chronicling the decline and fall of that theory. At the same time, it presents a prehistory of quantum chromodynamics (QCD), the quantum field theory of the strong interactions that was adopted in the 1970s. Complementing other prehistories of QCD that trace the development of attempts to describe the strong interactions (Cao, 2010), this Element will show how certain problematic aspects of the short-distance behavior of QFT came into the focus for theoretical physicists.

When QCD was shown to be asymptotically free in 1973, this killed two birds with one stone: it explained the presence of point-like constituents in nucleons, as observed in deep inelastic scattering experiments, and it demonstrated that the theory did not encounter the high-energy difficulties found in QED (Gross, 1990, p. 101). The first aspect was the crucial one in the discovery of asymptotic freedom and probably also in the acceptance of QCD as a theory of the strong interactions. But it was the latter aspect that ultimately made QCD the theoretical gold standard in QFT – less precise in its predictions than QED, but more likely to be a consistent theory. While a rigorous mathematical existence proof is still outstanding, it is no coincidence that the aforementioned Clay Millennium Problem explicitly demands such a proof for a (QCD-style) quantum Yang–Mills theory (rather than, say, an abelian gauge theory like QED).[7]

That would pretty much sum up the scope of this Element – as well as its temporal horizon, spanning from the development of renormalized QED in the late 1940s to the mid 1960s and the eve of the SLAC experiments that first hinted at asymptotic freedom – if it weren't for one thing: what exactly is meant by this word "consistency" that I keep bandying about? Historians might worry about whether it is an analytical category, designed to lump together a variety of historically distinct debates, or a category used by the historical actors themselves. This question can be answered rather easily: the reader will frequently encounter explicit references to consistency in historical quotes throughout this Element. Not everyone liked to use the concept of consistency; but it was used frequently enough that I can confidently state that the debate chronicled in this Element was widely regarded as a debate about the consistency of QED.

Philosophers, in turn, may worry that the historical actors invoked the concept of consistency somewhat too readily. I will be generous in this regard, and in general it will be possible to reconstruct the historical arguments as genuinely concerning questions of consistency. That is not to say that there was a universal or unchanging definition of (in)consistency underlying this debate. To the contrary. The historical development provides ample opportunity to analyze the varied notions of (in)consistency that were in play, and, toward the end, to reflect on their relevance and interrelations. This final focus of the Element will ultimately take us beyond the immediate concerns of the historical debate on the consistency of QED and allow us to draw broader philosophical lessons on the mathematical and logical structuring of knowledge in modern fundamental physics – arguably the most formalized of all the sciences.

But first, we return to Dyson. In Section 2, we will see him attempting to demonstrate the consistency of renormalized QED by showing that the known approximate solutions actually approximated exact ones – and ultimately we will see him fail, because of the divergence of the perturbation series.

---

[7] This change of paradigm is evident when comparing the Millennium Problem with a similar (though much longer) list compiled in the mid 1970s for a symposium of the American Mathematical Society. At the time, the central mathematical problem in QFT, as put forth by Arthur Wightman, was still to "prove the existence of a solution of the *quantum electrodynamics* of massive spin one-half particles in four-dimensional space-time" (Browder, 1976, p. 78, emphasis added). This quote highlights an additional advantage of Yang–Mills theory, as pointed out to me in conversation by Arthur Jaffe: the theory containing only gauge bosons is already nontrivial due to gauge boson self-interaction. By contrast, to get a nontrivial abelian gauge theory, one must introduce an additional charged matter field.

This theme – the impossibility of proving the consistency of QED – is then continued throughout the first volume of this Element.

After Dyson's failed attempt, it was clear that one had to go beyond perturbation theory in order to investigate the consistency of QED. In Section 3, I will look at the attempts to construct exact solutions in QED nonperturbatively. As we shall see, this goal ultimately became elusive, as the traditional dynamical equations of quantum theories (Schrödinger equation or Heisenberg equations of motion) were replaced in QFT by an infinite set of coupled integro-differential equations – the Schwinger–Dyson equations. At best one could investigate whether hypothetical nonperturbative solutions still contained the infinities that were removed through renormalization in the perturbative framework. In Section 4, we will discuss Gunnar Källén's proof that at least one of the renormalization constants was infinite also nonperturbatively. This issue was then connected to the finiteness of QED for short distances (or, equivalently, high – "ultraviolet" – energies) through the renormalization group methods developed by Murray Gell-Mann and Francis Low. With a construction of exact solutions to the Schwinger–Dyson equations out of the question, several physicists in the mid 1950s attempted to address the question of consistency on a different level by exploring the axiomatic foundations of QFT. I will discuss this approach in Section 5, focusing on the emergence of one of the central challenges for axiomatic QFT, Haag's theorem.

The second volume of this Element to be published in this series, is primarily concerned with the discovery of actual – though ultimately inconclusive – inconsistencies. At first such inconsistencies were discovered in simpler model QFTs, where exact solutions (of the Schrödinger equation) could still, in principle, be constructed. The discovery of pathological solutions in one such model QFT, the Lee Model, is the subject of Section 1 (of the second volume). Exact solutions for QED remained out of reach, but similar pathologies were discovered in approximate (though nonperturbative) solutions to the Schwinger–Dyson equations. This so-called Landau pole is discussed in Section 2.

By the late 1950s, a handful of central difficulties in proving the consistency of QED had thus been established: the divergence of the perturbation series, ultraviolet finiteness, Haag's theorem, and the Landau pole. However, the fundamental status of these difficulties remained disputed. We will look more closely at the ensuing debates, on the Landau pole (Section 3) and on ultraviolet finiteness (Section 4). In both cases, a consensus emerged that the original claims by Landau and Källén, respectively, had probably been overstated. Nonetheless, a final attempt in the mid 1960s by Kenneth Johnson, Marshall Baker, and Raymond Willey to prove the consistency of QED also ended in failure. Quantum electrodynamics appeared to be a lost cause; but it became clear through these investigations that there might actually be QFTs that fared better than QED, in particular with regard to their short-distance behavior. Kenneth Wilson's first explorations of this possibility will be examined in Section 5. We will see how Wilson formalized the study of the ultraviolet (UV) behavior of QFTs by introducing the concept of a running coupling constant, thereby setting the stage for the subsequent discovery of asymptotic freedom. Having thus reached the end of the story, Section 6 will present my reflections on the continued relevance of the consistency debate and the diverse attitudes toward inconsistency that we will encounter throughout this Element.

To arrive at these conclusions, we must venture deep into the quagmire that is the formalism of QFT, widely regarded as complex and recalcitrant; indeed, that complexity and recalcitrance is largely what this Element is about! I have aimed to make this Element accessible to readers who have attended an introductory course in QFT, have seen a field operator and the derivation of the Feynman rules. In fact, I even hope that this Element can also serve as an introduction of sorts to those more problematic aspects of QFT that are not usually addressed in introductory courses. And with that final disclaimer, we return to where this introduction began, the year 1949 and the brave new world of renormalized QED.

2 The Divergence of the Perturbation Series

2.1 Inconsistency in QED before Renormalization

Quantum electrodynamics is the relativistically invariant theory of electrons and positrons (represented by a fermionic Dirac field ψ) interacting with the electromagnetic field (represented by the four-vector potential A_μ).[8] The foundational equations of QED – the Lagrangian and the field commutation relations – were first written down in the late 1920s by Werner Heisenberg and Wolfgang Pauli, in the immediate aftermath of the development of quantum mechanics (Heisenberg and Pauli, 1929). Quantum mechanics primarily dealt with the energy levels of material systems, such as atoms; QED was supposed to extend it into a theory that could also treat transitions from one level to another, and the accompanying emission or absorption of electromagnetic radiation, in

---

[8] Using the terminology of the philosophy of science, it might be more accurate to refer to QED as a "model" rather than a "theory" – see, e.g., Koberinski (2021, p. S3749). I will, however, be adopting the terminology of physics, where "quantum field theory" refers to both the general framework and (sometimes using the plural, "quantum field theories") to specific models. I reserve the term "model" primarily for toy models, that is, unrealistic QFTs primarily studied for their formal properties.

a fully dynamical manner.[9] Quantum electrodynamics fulfilled these expectations well enough to first order in perturbation theory, that is, for the first term in a series expansion in powers of the fine structure constant $\alpha = e^2/\hbar c \approx 1/137$.

This first-order approximation was sufficient for most problems, in particular for calculating the probability amplitude for an atom to perform a transition. Second-order perturbation theory, in addition to providing small corrections to the results of first-order calculations, also introduced an effect of the radiation field on the energy levels of the atom itself. Here things started to go wrong. Quantum mechanics by itself provided very accurate predictions for the hydrogen spectrum and its fine structure. Instead of predicting small, hitherto unobserved displacements in this spectrum, calculations in QED predicted displacements of infinite magnitude (Oppenheimer, 1930).

There has been a fair amount of historical scholarship on this early period of quantum field theory up to and including the development of renormalization methods. My detailed historical reconstruction will thus only begin after the first successes of renormalized QED, around 1948. In this first section, I will instead take more of a bird's-eye view, addressing how the problem of the consistency of QFT was viewed in its first two decades. In this, I will refrain from making overly general statements about inconsistency in physical theory. Peter Vickers (2013, p. 15) has cogently argued that "some of the things we call 'theories' will have different characteristics, such that inconsistency means something different in each case." His method of analyzing inconsistency in science is not to engage with the structure of theories as a whole, but to identify a set of propositions relevant to the claimed inconsistency. Such a set of propositions can be equivalent to the entire theory, but it doesn't have to be. For QED, writing down such a relevant set of propositions is deceptively simple: it consists of (A) the basic tenets of quantum mechanics (as would have been found at the time in the textbooks by Dirac [1930] or von Neumann [1932]) combined with (B) the specific dynamical variables of QED and the dynamical equations they obey. In the Heisenberg picture, these were given by[10]

$$\left(i\gamma^\mu \partial_\mu - m\right)\psi(x) = e\gamma^\mu A_\mu(x)\psi(x)$$
$$\Box A_\mu(x) = e\overline{\psi}(x)\gamma_\mu \psi(x). \tag{1}$$

---

[9] On the role of radiation in the early years of quantum mechanics, see Blum and Jähnert (2024).
[10] The Heisenberg picture formulation of QED was first given by Heisenberg (1931) himself. Note that I have made one relatively small modification compared to the equations introduced there: I have written the Maxwell equations in the Lorenz gauge in terms of the potential, rather than in terms of the electric and magnetic fields.

From these propositions, Oppenheimer's result of infinite frequencies for the spectral lines could be deduced, along with similarly infinite results for other observable quantities. Alexander Rueger (1992) has mapped out several "Attitudes towards Infinities" adopted by physicists in the 1930s. He points toward Heisenberg's attempts at providing a physical interpretation of the infinities. But apart from these unsuccessful and unpublished attempts of the mid 1930s, there was a general consensus among physicists that the appearance of infinities in QED directly implied the theory's inconsistency.

Let us try to spell out the inference from the calculated infinities to the inconsistency of QED more explicitly. One might simply interpret it as an inconsistency between QED and empirical facts – after all, the frequencies of the spectral lines of hydrogen had been measured with definite, finite values. In this case, the inconsistency referred to by the historical actors would amount to nothing more than empirical inadequacy. However, the real problem was not that the infinities predicted by QED *did* not reproduce the observed spectra, but that they *could* not. A spectral line of infinite frequency was considered a contradiction in and of itself, without recourse to the frequency actually measured.[11] Physicists were thus implicitly invoking a more general meta-theoretical criterion that did not rely on empirical facts.

What should we take this implicit criterion to be? A first attempt might be to add the claim (C) "observable quantities do not take infinite values" to the initial set of propositions. But this seems a bit too specific and tailored to the situation at hand. Instead, I suggest that, in order to spell out the argument for inconsistency, we should add the more general proposition (C') "the dynamical equations have solutions" to our original set.[12] Consider two points: (i) the solutions obtained to the dynamical equations of QED did not simply diverge for certain values of parameters, for high energies, say (a domain that will become important later on) – rather one obtained results that were universally infinite; (ii) there was no obvious way to use these infinities to make QED predict *arbitrary* values, for example, for the radiative corrections to the hydrogen spectrum. Instead, QED very specifically predicted infinite corrections. It thus seems fair to say that the dynamical equations of QED simply had no solutions at all, just as we would say that the following set of algebraic equations

---

[11] In this regard, the situation was similar to the UV catastrophe in blackbody radiation, where Einstein (1905, p. 136) had remarked that the Rayleigh–Jeans law "not only fails to agree with experience, but it also states that in our model a definite distribution of energy between ether and matter is out of the question."

[12] One might consider further specifying the domain in which these solutions should lie. But physicists at the time were certainly not very explicit about this, so it seems reasonable to be as permissive as possible in this regard.

$$x + 2 = y + 1$$
$$x = y \qquad (2)$$

has no solutions and is thus *inconsistent* – even though $x = y = \infty$ is a solution of sorts. Adding a proposition such as (C') when evaluating the consistency of physical theories is not unprecedented. In his exploratory attempts at an axiomatization of mechanics, David Hilbert – arguably *the* authority on questions of mathematical consistency in the early twentieth century – implicitly used the existence of solutions to the dynamical equations as a consistency criterion (Majer, 2001, p. 23). Proposition (C') also seems in keeping with the practice of physics, which is far more concerned with solving equations than with logical deduction, ever since Leonhard Euler, in his reformulation of Newtonian mechanics, introduced "a new style of mathematical physics in which fundamental equations take the place of fundamental principles" (Darrigol, 2005, p. 25).[13] And we will see later on that the existence of solutions remained the relevant criterion when Dyson began to investigate the consistency of renormalized QED.

But how robust was the claim that the equations of QED have no solutions? So much is made and has been made of the infinities that appeared in early QED that it seems important to point out that these infinities merely appeared in the higher orders of a specific approximation method – perturbation theory. Why did physicists in the 1930s equate the breakdown of an approximation method with the breakdown of the entire theory? A prominent explanation was given by Niels Bohr:

> Their [Heisenberg and Pauli's] formalism leads, in fact, to consequences inconsistent with ... the small coupling between atoms and electromagnetic radiation fields, on which rests the interpretation of the empirical evidence regarding spectra, based on the idea of stationary states. Under these circumstances, we are strongly reminded that the whole attack on atomic problems leaning on the correspondence argument is an *essentially approximative procedure* made possible only by the smallness of the ratio between the square of the elementary unit of electric charge and the product of the velocity of light and the quantum of action which allows us to a large extent to avoid the difficulties of relativistic quantum mechanics in considering the behavior of extra-nuclear electrons. ... This is a non-dimensional constant fundamental for our whole picture of atomic phenomena, the theoretical derivation of

---

[13] A similar development can be observed in the history of electrodynamics, where Heinrich Hertz – who cast the theory in the compact form that was later printed on t-shirts – famously asserted: "This, and not Maxwell's peculiar conceptions or methods, would I designate as 'Maxwell's Theory.' To the question, 'What is Maxwell's theory?' I know of no shorter or more definite answer than the following: – Maxwell's theory is Maxwell's system of equations" (Hertz, 1893, p. 21).

which has been the object of much interesting speculation. Although we must expect that the determination of these constants will be an integral part of a general consistent theory in which the existence of the elementary electric particles and the existence of the quantum of action are both naturally incorporated, these problems would appear to be out of reach of the present formalism of quantum theory, in which the complete independence of these two fundamental aspects of atomicity is an essential assumption. (Bohr, 1932, pp. 378–379)

From Bohr's characteristic verbosity we can extract the following argument: it is not at all surprising that QED breaks down at second order in perturbation theory. First-order perturbation theory, based on the "small coupling between atoms and electromagnetic radiation fields," is not just some arbitrary approximation – it is an essential foundation that we rely on every time we calculate the energy of a stationary state without worrying about radiation.

To better understand this perspective, let us look back all the way to Bohr's atomic model. It makes two claims about radiation: (a) electrons in a stationary state do not emit radiation, and (b) monochromatic radiation is emitted in discrete quantum jumps from one stationary state to another. In hindsight, these are not really fundamental postulates of quantum theory. Rather, they are statements about the validity of first-order perturbation theory. Claim (a) simply states that energy levels are not corrected until second order in perturbation theory, while (b) just states that transitions in first-order perturbation theory only involve *one* electromagnetic field operator, and thus only *one* photon with a single radiation frequency. This did not change in quantum mechanics, where Bohr's second postulate was replaced by Fermi's golden rule, which can be obtained directly from first-order perturbation theory.

Bohr's view was that all of quantum mechanics was an "essentially approximative procedure," a fact that was made manifest by the divergences appearing in second-order perturbative QED. This would only be resolved within a "general consistent theory," in which one would be able to predict the value of the fine structure constant and thereby relate the elementary unit of charge e to Planck's constant.[14]

In Bohr's view, there was thus no point in looking for nonperturbative solutions. Quantum electrodynamics was built on quantum mechanics and quantum mechanics was built on perturbation theory; perturbation theory was thus hardwired into the structure of QED. The breakdown of QED perturbation theory at second order thus implied the breakdown of QED *tout court*. In fact, it even implied the breakdown of quantum mechanics itself.

---

[14] Attempting to derive the value of the fine structure constant – or at least anticipating its derivability – was very popular at the time, see Kragh (2003).

Alexei Kojevnikov has emphasized that attitudes toward theory change in the early 1930s were shaped (one might also say: distorted) by the radical developments of the preceding decades, and that physicists thus eagerly anticipated yet another conceptual revolution (Kojevnikov likens this to Trotsky's notion of "permanent revolution"), rather than looking for more immediate solutions to the theoretical anomalies they encountered.[15] A similar point was made by Carl Friedrich von Weizsäcker in his reflections on his mentor Heisenberg:

> In 1912, Bohr's fundamental insight had been that explaining the stability of atoms would require not just a new atomic model, but new fundamental laws of physics. In 1925, Heisenberg had, at age 23, given these laws – under the name of quantum mechanics – their final shape, which they have kept to this day. On the level of philosophical reflection, this became for him the paradigmatic example of the idea that fundamental theoretical physics does not proceed in the steady accumulation of knowledge, but rather in a sequence of closed theories. Neither he nor Bohr then considered quantum mechanics to be the last fundamental theory; they expected imminent advances beyond this step. Heisenberg could never bear the thought that his youthful work should have been the last great inspiration of his life. (von Weizsäcker, 1999, p. 294)[16]

And Alexander Rueger (1992, p. 315) remarks that "the strong emphasis on deficiencies pointed to the framework of expectations in which the authors worked." It thus seems fair to say that many of the protagonists of early QFT were overly quick to accept the conclusion that QED was inconsistent, because they anticipated, from the start, that some "future final theory" (Heisenberg and Pauli, 1929, p. 3) would replace it – and all of quantum mechanics along with it.

However, one did not need to hold such strong views on the future development of physics in order to equate the breakdown of perturbation theory with the inconsistency of QED. Based on a power series expansion in a very small parameter, perturbation theory was prima facie a very good approximation, as emphasized by the Swedish physicist Ivar Waller (1930, p. 675):

> This calculation of course leaves open the possibility that the mistake is in the approximation method. But our method of solution seems to conform to the spirit of the theory, since the charge could in principle be taken to be arbitrarily small. More generally, a different method of solution hardly seems possible.

---

[15] Talk at the HQ4 conference in San Sebastián (2015), www.youtube.com/watch?v=Np2LGUyYZa8).

[16] I would like to thank Bernd Henschenmacher for pointing me to this passage.

We can see here how quick physicists were to move from the in-practice unsolvability, as seen in the divergences arising in perturbation theory, to asserting in-principle unsolvability. But this was a crucial step. In-practice unsolvability would have been nothing new in physics. A famous example is hydrodynamics: proving that the Navier–Stokes equation has exact solutions ranks alongside the existence of Yang–Mills theory as one of the Clay Institute Millennium Problems. Yet, despite the longstanding difficulties involved in solving the dynamical equations of hydrodynamics, these were never viewed as implying the inconsistency of theoretical hydrodynamics. Although Leonhard Euler "deplored the difficulties of solving his [hydrodynamical] equations" (Darrigol, 2005, p. 25), he stated his conviction that

> If it is not permitted to us to penetrate to a complete knowledge concerning the motion of fluids, it is not to mechanics, or to the insufficiency of the known principles of motion, that we must attribute the cause. It is analysis itself which abandons us here since all the theory of the motion of fluids has just been reduced to the solution of analytic formulas. (Truesdell, 1954, pp. LXXXVI–LXXXVII)

This sentiment – that the unsolvability of the equations of hydrodynamics was a mathematical difficulty that did not affect the evaluation of hydrodynamics as a physical theory – was frequently echoed over the decades.[17] Indeed, outside the immediate context of hydrodynamics, one can rather observe the *reverse* argument: namely, that the physical cogency of an equation guarantees its solvability. This was notably emphasized by the French mathematician Jacques Hadamard:[18]

---

[17] We find, e.g., Hermann von Helmholtz (1873, p. 502) remarking that "there is indeed, as far as I can see, currently no reason not to consider the hydrodynamical equations as the exact expression of the laws really governing the motions of fluids," while immediately pointing out that the Navier–Stokes equation, which describes the motion of a viscous fluid, could not be solved. Or, as John Wheeler would put it some 80 years later: "There are no rivers on the earth because the eq[uatio]ns of hydro[dynamics] are too complicated to solve" (Relativity Notebook III, p. 130, entry dated March 27, 1955, JWP).

[18] Another example of this view can be found in Felix Klein's assessment of Bernhard Riemann's existence proofs for solutions of the Laplace equation with given boundary conditions. In the 1850s, Riemann was working on the theory of complex-differentiable functions, whose real and imaginary parts satisfy the Laplace equation in two dimensions. When proving the existence of solutions, Riemann relied on what would come to be known as Dirichlet's principle. This method had been introduced by Riemann's teacher, Peter Gustav Lejeune Dirichlet, when proving the existence of solutions to the Laplace equation in the physical context of gravitational and electrostatic potentials. It starts from the set of integrals

$$U = \int_\Omega \left(\left(\frac{\partial u}{\partial x}\right)^2 + \left(\frac{\partial u}{\partial y}\right)^2 + \left(\frac{\partial u}{\partial z}\right)^2 \right) dV$$

for all functions u in a volume Ω that fulfill the given boundary conditions (though not necessarily the Laplace equation). Dirichlet's principle then states that there exists a function u

> It is remarkable... that a sure guide is found in physical interpretation: an analytical problem always being correctly set, in our use of the phrase,[19] when it is the translation of some mechanical or physical question. (Hadamard, 1923, p. 32)

Physicists in the 1930s did not have this sort of confidence when it came to QED. I argue that this was because, rather than having physical reasons to expect the existence of solutions, they had a ready physical interpretation for the unsolvability of QED – an interpretation also put forward by Bohr in his previously cited Faraday lecture:

> The scope of the quantum mechanical symbolism is essentially confined, however, to problems where the intrinsic stability of the elementary electrical particles can be left out of consideration in a similar way as in the classical electron theory. In this connexion, it must not be forgotten that the existence of the electron even in classical theory imposes an essential limitation on the applicability of the mechanical and electromagnetic concepts. ... The difficulties inherent in any symbolism resting on the idealisation of the electron as a charged material point appear also most instructively in the recent attempt of Heisenberg and Pauli to build up a theory of electromagnetic fields on the lines of quantum mechanics. (Bohr, 1932, pp. 377–378)

Bohr thus argued that the inconsistency observed in QED was a result of transgressing the theory's domain of applicability – a domain of applicability it had largely inherited from classical electron theory, the result of an idealized description of charged matter as point particles. In classical electron theory, the domain of applicability was defined by the negligibility of radiation reaction, that is, the force experienced by an electron due to the field it generates itself.[20]

---

that minimizes U. One could then easily show that this minimizing function also fulfills the Laplace equation, completing the existence proof.

Riemann was heavily criticized for his use of Dirichlet's principle by Karl Weierstrass, known as a paragon of rigor with little interest in physics. Weierstrass emphasized that one could only state with confidence that the set of integrals U has a greatest lower bound, not that there exists a function u for which U actually takes a minimal value. As Riemann's results gradually came to be corroborated through other methods after his death, Klein asserted that, in using Dirichlet's principle, Riemann had merely been trying to "support the physical evidence through a mathematical deduction" (Klein, 1882, p. IV).

For more on the history of Dirichlet's principle, see Monna (1975). The contrast between the physically intuitive Riemann and the mathematically rigorous Weierstrass was a central theme in Klein's later work on the history of mathematics (Klein, 1926, chapter VI), and we will encounter this trope again soon.

[19] "Correctly set" is a translation of Hadamard's French expression *correctement posé*, which he used more or less interchangeably with *bien posé* (Hadamard, 1907). I have not been able to ascertain when the now-current English expression "well-posed" came into use. I would like to thank Gordon Belot and Dominic Dold for discussions on this point.

[20] For an excellent discussion of this subject, which significantly informs the following analysis, see Vickers, (2013), chapter 4 and section 7.4.

As soon as radiation reaction had to be taken into account – e.g., for relativistic electrons or electrons in a synchrotron – one also needed to take into account the structure of the electron. However, no viable models of the structure of the electron were known: point particles led to infinitely large field energies close to the particle, while extended particles could not be kept stable, as there was no way of reconciling the necessary cohesive forces with special relativity.

In the classical theory, these difficulties could be avoided by simply neglecting the radiation reaction. All one had to do was eliminate the electron's interaction with its self-field from the Lorentz force equation. The electron could then be treated as an idealized point charge. The price one had to pay was that one, more or less voluntarily, brought in a new form of inconsistency: energy is not conserved, because the field that appears in Maxwell's equations (the total field) was not identical to the field that appears in the equations of motion of the electron (the external field, with the electron's self-field neglected). This is the "inconsistency of classical electrodynamics" highlighted by Mathias Frisch (2005).

The divergence difficulties of QED appeared to be analogous to the classical situation, because higher orders of perturbation theory could be physically interpreted as radiation reaction. The shift in energy levels and spectral lines due to radiative corrections clearly seemed to be the result of the electron's interaction with its self-field. Unlike the classical case, however, that self-field was not to be identified with radiation actually emitted, but was instead purely "virtual." This problematic aspect of the analogy did lead some physicists to view the inconsistency of QED as a genuine quantum phenomenon, arising from the presence of a continuous infinity of virtual states (Blum, 2017, pp. 27–28, 41–42). But, by and large, the analogy between the divergences of QED and the classical problem of radiation reaction was accepted.

The analogy was extended by equating the local interactions of QED – that is, the evaluation of all fields in the dynamical equations at the same space-time point – with the assumption of point-like electrons in the classical theory.[21] The divergences resulting from the local interactions could be avoided by introducing a high-energy cutoff into QED. This was the analog of using an extended electron in the classical theory. But both of these solutions appeared impossible to reconcile with special relativity (Blum, 2017, section 1.1).

What, then, was the quantum analog of the classical procedure of simply neglecting radiation reaction? In quantum theory one could achieve essentially the same thing by restricting oneself to leading-order perturbation theory. And with this restriction, QED was actually rather successful empirically, providing

---

[21] See, for example, Waller (1930, p.676).

good quantitative descriptions of processes such as the emission and absorption of radiation (Dirac, 1927b), optical dispersion (Dirac, 1927a), or Compton scattering (Klein and Nishina, 1929). And if one simply ignored the infinities, the Dirac equation – unmodified by any QED corrections – gave excellent predictions for the fine structure of hydrogen (Dirac, 1928a,b). For a time, some physicists endorsed such an approach to radiation phenomena, labeling it the "correspondence" approach because it preserved the close formal connection to classical theory that had been so essential in the construction of quantum mechanics. This idea was first proposed by Heisenberg (1931), but, ultimately, it did not catch on.[22] In an interview with Thomas Kuhn and John Heilbron,[23] Léon Rosenfeld recalled the strong resistance this approach encountered:

> I gave a paper on that at the congress in Rome [in 1931 (Rosenfeld, 1932)], but it fell completely flat. People started saying, "Yes, but this is no theory," or asking questions, "You have the Hamiltonian; why do you forbid us to handle the Hamiltonian according to the recipes?"... I said then finally, 'Yes, but what I intended to show is that by a simple correspondence prescription one can get all safe results... and the suggestion is that the only way one can work with this formalism is by not considering it a closed formal scheme." Then Fermi, who was in the chair, just to close the discussion, said, "Yes, we have understood that, but that is not the usual conception one has of physical theories." That was the general feeling.

Just as in the classical theory, radiation reaction could only be removed by introducing a new kind of inconsistency. The original dynamical equations of QED (equation 1) were a "closed formal scheme": the goal was to find (operators) A_μ and ψ that, when inserted into the respective right-hand sides, yielded equations whose solutions were again the A_μ and ψ one started out with.[24] Aborting the approximation meant breaking this closed structure: the expressions one inserted no longer had to be *exactly* the expressions one got out. And physicists did not consider this option more attractive than the divergences of higher-order perturbation theory. As Rosenfeld went on to remark in the aforementioned interview with Kuhn and Heilbron, the correspondence approach was viewed as "the extreme position of defeat."

In the 1930s, the status of QED was thus fundamentally unclear. There were actual textbooks written on the subject in the 1930s and 1940s, but they included disclaimers such as the following:

---

[22] See also (Rueger, 1992, pp. 317–322).
[23] Session II, July 19, 1963. www.aip.org/history-programs/niels-bohr-library/oral-histories/4847-2.
[24] Of course, there are significant subtleties involved here, in particular when multiplying two operator fields defined at the same space-time point. We will discuss these difficulties in due time, when they become relevant historically.

> At first it seemed that... a consistent quantum theory of the electromagnetic field could not be found.... It seems now that there is a certain limited field within which the present quantum electrodynamics is *correct*... The present theory can be correctly applied to the interaction of light with elementary particles in the first approximation. The difficulties which occur, especially in the higher approximations to this interaction,... show the limits within which the theory is valid.... If the application of the theory is confined within these limits, it will be seen in this book that the theory gives – qualitatively and quantitatively – a full account of the experimental facts within a large field... Thus it seems that the theory is well enough developed, and the limits of application well enough marked, for a summary to be given.[25]

As long as QED, aborted after the first approximation, gave perfect predictions within its very own domain (the emission and absorption of spectral radiation), the inconsistencies remained strangely inconsequential. This changed only when high-precision measurements indicated that the second approximation might be physically relevant after all.

2.2 Oppenheimer and Dyson

The first discrepancies started to appear in the late 1930s, but remained controversial. It was not until 1947 that Willis Lamb, building on his experience with WWII microwave radar technology, observed for the first time induced transitions from the $2S_{1/2}$ to the $2P_{3/2}$ levels of hydrogen and found the transition frequency to be about 1 GHz smaller than expected. This "Lamb shift" soon came to be understood as the result of radiative QED corrections, implying that these corrections were neither infinite nor zero, but rather measurable and small. This led, almost immediately, to the development of renormalization methods, enabling the calculation of these corrections, and putting QED on track toward becoming the ultra-high precision theory it is now celebrated as.[26]

In the early years, these renormalization methods were very much in flux and far from standardized. Paul Matthews and Abdus Salam famously remarked that in renormalization theory, "[t]he difficulty [...] is to find a notation which is both concise and intelligible to at least two people of whom one may be the author" (Matthews and Salam, 1951, p. 314). Throughout this Element, I will

---

[25] (Heitler, 1936, pp. ix–x), emphasis in the original.
[26] The standard reference on this development is, of course, (Schweber, 1994). In this very brief summary, I have focused exclusively on the Lamb shift; an essentially parallel story can be told for the anomalous magnetic moment of the electron. I have also not addressed the theoretical developments of the 1930s and 40s that partially prepared the way for covariant renormalization methods, not least because I have argued elsewhere that these developments have frequently been overestimated in the past (Blum, 2015).

homogenize conventions from the historical sources to keep the ideas we are interested in from being drowned in elaborations on notational differences and historical minutiae.[27] I have thus chosen to base my very brief sketch of renormalization theory, which primarily serves to introduce terms and conventions that will reappear throughout the Element, on a somewhat later review article by Matthews and Salam (1954). This article already contains elements not found in Schwinger's and Feynman's original tool kits of the late 1940s. Nonetheless, I believe that it is close enough to the spirit of these early works, while simultaneously highlighting novel aspects that will become important only in later sections.

The first step in renormalization was to brand the fields and parameters that appear in the original (now: unrenormalized) QED field equations as unobservable "bare" fields and parameters, earning them a zero subscript as in m_0 for the bare mass.[28] One then performed a transformation from these unrenormalized quantities to the (observable) renormalized fields and parameters, which were denoted by the original symbols, without a subscript. The transformation to the renormalized field equations (and to the associated equal-time commutation relations) involved the following four substitutions:[29]

$$m_0 \to m - \delta m$$
$$\psi_0 \to \sqrt{Z_2}\psi$$
$$A_0^\mu \to \sqrt{Z_3}A^\mu$$
$$e_0 \to \frac{e}{\sqrt{Z_3}}$$

---

[27] Unfortunately, Matthews and Salam's quip remains true to some extent, more than seven decades later. There are many different notational conventions around, none of which are entirely satisfactory. In particular, there is no accepted notational convention that distinguishes between free, unrenormalized, and renormalized quantities in a consistent manner. I considered introducing a more unified notation but ultimately decided that a history book was not the right place to do that. Therefore, in the convention used here, we also have, e.g., the absence of an additional subscript implying a renormalized quantity when dealing with fields and parameters (ϕ, m) but implying a free quantity when dealing with propagators (D_F).

[28] The term "bare" is somewhat ambiguous. "Bare" generally means "without radiative corrections included," but this can refer to two distinct concepts. On the one hand, it can refer to the inclusion of radiative corrections through renormalization. This is the sense in which I will be using the term throughout this book; "bare" is then simply a synonym for "unrenormalized," see, e.g., (Weinberg, 1995, p. 472). But "bare" can also, somewhat less frequently, refer to the inclusion of radiative corrections by adding higher-order perturbative terms; thus (see, e.g., (Fried, 1959) or (Carrington et al., 1999)) the "bare propagator" is D_F. I will not be using "bare" in this latter sense and will refer to D_F only as the free propagator.

[29] These transformations have already been simplified by using gauge invariance and the resulting Ward identity.

Here, δm is the mass renormalization constant, Z_2 the (electron) wave function renormalization constant, and Z_3 the charge (or photon wave function) renormalization constant. These constants can be chosen so as to cancel, term by term, the divergences that appear when perturbatively constructing scattering amplitudes from the field equations.[30] The renormalized fields and parameters were then entirely free of divergences and were identified with the particles and constants actually measured in experiment – the "physical" particles and constants, as they came to be called, in contradistinction to their "bare" counterparts. And that was it – pretty simple in theory, but rather complicated in practice, especially at higher orders of perturbation theory.

By the summer of 1948, it was clear that the results of the new precision measurements could be reproduced using renormalized QED. At the Eighth Solvay Conference in Brussels, in the fall of 1948 (several months before his debate with Dyson), Oppenheimer (1950) delivered a report on "Electron Theory," in which he outlined the current status of QED.[31] He expressed his conviction that renormalization had not turned QED into a "completed consistent theory" (p. 272) – almost, but not quite. Renormalization only served to "isolate, recognize and postpone the consideration of those quantities ... for which the present theory gives infinite results" (p. 273). This "absence of complete closure" (p. 273) could no longer be blamed on the neglect of radiation reaction, which had, after all, become observable and calculable in the Lamb shift. Instead, Oppenheimer connected it with the neglect of the nuclear interactions, which had emerged as the new frontier of fundamental physics over the preceding two decades:

> [F]or mesons and nucleons generally, we are in a quite new world, where the special features of almost complete closure that characterizes electrodynamics is quite absent. That electrodynamics is ... not quite closed is indicated, not alone by the fact that for finite $e^2/\hbar c$ the present theory is not after all-consistent [sic], but equally by the existence of those small interactions

---

[30] Investigating the residual arbitrariness involved in correctly choosing the renormalization constants was one path that led to the discovery of the renormalization group (Stueckelberg and Petermann, 1953). In this Element, in Section 4, we will only consider the path taken by Gell-Mann and Low, which instead went through the investigation of short-distance behavior. On the two paths to the renormalization group, see Fraser (2021).

[31] The only source we have for Oppenheimer's report is the version published in the proceedings, which appeared two years after the conference. Oppenheimer clearly revised his original manuscript, since the published version refers to Dyson's S-Matrix paper, which was submitted on February 24 and published on June 1, 1949, as "in press" (fn. 33). However, this reception of Dyson's work would, if anything, have softened Oppenheimer's negative attitude toward QED, which I emphasize here and which still shines through clearly in the published report.

with other forms of matter to which we must in the end look for a clue, both for consistency, and for the actual value of the electron's charge (p. 281).[32]

As opposed to radiation reaction, the nuclear interactions were not something that QED could in any way be considered responsible for. There was thus no point in looking to QED for further insights: "[T]he structure of the theory itself gives no indication of a field strength, a maximum frequency or minimal length, beyond which it can no longer consistently be supposed to apply" (p. 272). Oppenheimer therefore looked to the experimental study of the nuclear interactions for entirely novel theoretical insights, independent of the theoretical structure of QED.

He was aware that this was not the only option. The idea that QED might serve as an inspiration, a blueprint even, for a theory of the nuclear interactions had first been introduced by Hideki Yukawa in the mid-1930s and had never really gone away. In the discussion after his talk, Oppenheimer acknowledged that he himself had "at one time believed" that "the real importance" of renormalization "lay in the fact that one had an entirely new way of dealing with the Maxwell–Yukawa analogy" (p. 284). And while he now appeared to believe that "this analogy is rubbish," he conceded that a more in-depth understanding of renormalized QED might be worth pursuing. Not in order to see "whether one can calculate to one part to 10^8 th [sic], the Lamb shift," but in order to explore whether a theory of the nuclear interactions could be modeled after QED, "even if one thinks the results would be negative" (p. 284).

When Oppenheimer returned from Europe to the IAS after the Solvay conference, Dyson had already begun his one-year fellowship at the institute. Dyson had just submitted a major paper on QED, in which he showed that the various covariant approaches to QED and renormalization, by Julian Schwinger, Richard Feynman, and Sin-Itiro Tomonaga, were, in fact, equivalent (Dyson, 1949a). But, as we have seen, Oppenheimer was not particularly invested in the further development of renormalized QED. When approached by Dyson, he simply handed him a copy of the Solvay talk. In response, Dyson drafted a short memorandum outlining his views on QFT, which he sent (apparently by mail!) to Oppenheimer on October 18, 1948 and in which he asserted:[33]

---

[32] Here we see Oppenheimer hinting at the hope that a final theory would also determine the value of the fine structure constant, see footnote 14.

[33] This short summary of the first interactions between Oppenheimer and Dyson is based on Schweber, (1994, pp. 520ff), where the reader will also find the full text of Dyson's memorandum. Dyson (1979, chapter 7) also presented his own recollections, which are more impressionistic (e.g., not mentioning the memorandum), but do contain some memorable remarks (p. 73) on his clash with Oppenheimer: "[Oppenheimer] had somehow become convinced during his stay in Europe that physics was in need of radically new ideas, that this quantum electrodynamics of Schwinger and Feynman was just another misguided attempt

I do not see any reason for supposing the Feynman method to be less applicable to meson theory than to electrodynamics. In particular I find the argument about "open" and "closed" systems of fields irrelevant.

Dyson's program stood in stark contrast to Oppenheimer's – QED was supposed to act as a role model for future field theories of the nuclear interactions. And Dyson knew what the next step needed to be in order to show – and convince Oppenheimer – that this was indeed a viable approach:

> What annoyed me in Oppenheimer's initial lethargy was not that my finished work was unappreciated, but that he was making it difficult for me or anybody else to go ahead with it. What I want to do now is to get some large-scale calculations done to apply the theory to nuclear problems, and this is too big a job for me to tackle alone. So I had to begin by selling the theory to him. As soon as he understands and believes in it, he will certainly have a great deal of useful advice and experience to offer us in applying it. Also he may be able to help me decide what I should do next, though *I am fairly determined already on a thoroughgoing attempt to prove the whole theory consistent.*[34]

Over the next couple of years, Dyson thus made it his mission to show that QED was a consistent theory and could serve as a a starting point for the study of the nucleus. Within the next couple of months, he famously proved that the S-Matrix in QED could be made finite to *all* orders of perturbation theory through renormalization (Dyson, 1949b). This achievement sufficiently softened Oppenheimer's stance, allowing the two men to amicably agree to disagree at the January 1949 meeting that opened this book.

2.3 Investigating the Consistency of Renormalized QED

Dyson spent the rest of his year in Princeton enjoying the acclaim brought by his renormalizability proof and meeting his first wife, the Swiss mathematician Verena Haefeli, whom he married the following year. In the fall of 1949, he returned to England, where he spent the next two years at the University of Birmingham with Rudolf Peierls. During his first year at Birmingham, Dyson "was always dividing [his] time between five or six problems and never sat down and concentrated upon one thing long enough to finish it."[35] It was not until December 1950, after another extended stay at the IAS, that he finally

---

to patch up old ideas with fancy mathematics. I was delighted to hear him talk in this style.... Instead of arguing with Oppenheimer about the dubious merits of my own work, I would be fighting for the entire program of quantum electrodynamics... Instead of fussing over details, we would be clashing on basic issues. Already I could feel that the Lord had delivered him into my hands."

[34] Letter from Dyson to his parents, November 1, 1948 (Dyson, 2018, p. 113), my emphasis.

[35] Letter to Rudolf Peierls, August 3, 1950 (Lee, 2009, p. 244).

wrote a paper in which he laid out the steps still required to show that "quantum electrodynamics, in spite of its inherent divergences, constitutes a consistent and meaningful theory" (Dyson, 1951a, p. 428).

The two-step program Dyson laid out conforms with the suggested reading of "consistency" as "existence of solutions to the dynamical equations." The first step was to extend his renormalizability proof beyond the S-Matrix. While some physicists had conjectured that relativistic quantum theory merely consisted in determining the S-Matrix,[36] it was generally viewed as a derived quantity. The immediate solutions to the dynamical equations of QED (eq. 1) were the field operators ψ and ϕ. Dyson thus aimed to show that renormalization methods could also be applied to the (Heisenberg picture) field operators, thereby explicitly delivering solutions to the dynamical equations, solutions that were finite to all orders of perturbation theory. Dyson resolved the first point quickly and successfully, and we shall not discuss it much further. I will only mention that the methods he developed for this purpose turned out to be not as widely useful as his diagrammatic method for proving the renormalizability of the S-Matrix had been. This was a great disappointment to Dyson, who had hoped that investigating the consistency of QED would directly inform a QFT treatment of the nuclear interactions. As he later remarked in an interview with Sam Schweber: "This was not philosophically driven at all: it was intended as a practical tool" (Schweber, 1994, Section 9.16).

The second step, which we will now discuss in more detail, was to lift the restriction to perturbation theory. After all, not only were all attempts at finding solutions to QED still based on perturbation theory, the entire renormalization procedure itself relied on removing infinities term by term from a perturbation series. In December 1950, Dyson sent a copy of his new paper to Wolfgang Pauli, whom he regarded as "the one member of the 'old gang' who takes the trouble to thoroughly understand the new methods."[37] Pauli immediately seized on the restriction to perturbation theory as the central challenge for Dyson's program:

> I have no doubt that your good swimming belt will also be sufficient to drive you until the nth approximation regarding the renormalization of the fields as it is announced by you. It is only after this step that things will get interesting. You stirred the curiosity of the reader by your remark "incidentally it will appear" (that is a really nice way to put it) "that our methods provide a basis for removing defect (i) (namely the expansion of everything in power [sic] of the coupling constant e) also from the analysis".

---

[36] This is the view of Heisenberg's 1940s S-Matrix program, which Dyson viewed as closely related to Feynman's version of QED (Blum, 2017).

[37] Letter to Rudolf Peierls, March 31, 1949 (Lee, 2009, p. 179).

> Everything depends on whether you will get farer [sic] than Nr. II of the "series". This would be a really new step which seems to me in any case beyond the reach of your present swimming belt.[38]

This marked the beginning of a discussion on what it would actually mean to remove the restriction to perturbation theory. Dyson hoped to demonstrate the convergence of the perturbation series, which would then show that the approximative solutions one worked with corresponded to exact solutions of the dynamical equations of QED. Pauli, however, found this to be insufficient for two reasons: (i) "it would be practically useless in the case of a strong coupling (high values of the coupling constant)" (as might be expected for the nuclear interactions), and (ii) even a converging perturbation series provided little "physical enlightenment."[39] Pauli would later speak of the "veil of the perturbation series."

In his reply, Dyson reiterated the standpoint he had already defended with Oppenheimer two years earlier – proving the consistency of QED was not about gaining foundational insight, but about making immediate progress with quantum field theories of the nuclear interactions:

> The test which I wish to be applied to my methods is "Do they enable us to decide definitely the correctness or incorrectness of meson theories by a quantitative comparison with experiment?" If they can do this, I shall be very well satisfied.
> You are asking "Do the new methods give us any new theoretical insight into the foundations of physics?" This question I am content to answer in the negative.
> Of course, I too would like to escape altogether from series expansions, if I could do it. But I believe the idea of renormalization is linked in a very complete way, perhaps inseparably, with the series expansion.[40]

However, Pauli felt that a proof of consistency that was based on a perturbation expansion was not worth all that much. In his reply on February 18, 1951, he consequently objected:

> That "the idea of renormalization is linked in a very complete way, perhaps inseparably with the (power) series expansion" is just the point which makes me feel rather critically to the present quantized field-theory (Even if this power series is mathematically convergent and its convergence can be proved and in spite of the practical successes of the new quantum electrodynamics)... Even apart from practical questions of computation I have the

---

[38] Letter from Pauli to Dyson, December 20, 1950 (von Meyenn, 1996, p. 220).
[39] Letter from Pauli to Dyson, February 5, 1951 (von Meyenn, 1996, p. 262).
[40] Letter from Dyson to Pauli, February 15, 1951 (von Meyenn, 1996, p. 263).

definite impression that (also in electrodynamics) a final form of a fundamental law of nature should *not* be formulated in such a way that a power series development is *essential* to get the very logical (mathematical) meaning (definition) of this law.[41]

It is important to note that Pauli was not merely pitting his desire for physical insight against Dyson's mathematical rigor. He associated Dyson's perturbative approach with a *specific* mathematical tradition that he rejected, while simultaneously emphasizing, in the postscript to the letter cited earlier, that the two of them shared a much closer relationship with mathematics than most physicists. After all, Dyson had majored in mathematics as an undergraduate at Cambridge and had written several papers on number theory:

> P.S. I hope to have once a chat with you on the general relation of mathematics and physics. In your present age I believed that my talent is lying in a very mathematical direction, although in a kind of "applied mathematics," because my interest for the laws of nature was always very prominent. But then I fell into this spectroscopic "term-zoology" where my knowledge of the theory of complex analytic functions could not be used.
> The development of the big mathematical edifice of quantum-mechanics was done by others, but anyhow I could apply my knowledge of Cayley and F. Klein with this "spinors" [...] By the way, F. Klein was always very much against the overemphasizing of the power series (Weierstrass).

Mathematically, Pauli thus saw Dyson in the tradition of the nineteenth-century German mathematician Karl Weierstrass. In Felix Klein's "Vorlesungen über die Entwicklung der Mathematik im 19. Jahrhundert" (*Lectures on the Development of Mathematics in the 19th Century*), Klein had portrayed Weierstrass as the proponent of the brute-force application of power series, remarking, for example, that Weierstrass had calculated series coefficients "up to the 20th order with all of the, downright uncanny, numerical factors" (Klein, 1926, p. 280). Klein contrasted Weierstrass with Riemann, for whom power series had been merely "an occasional resource for the elaboration of his thoughts," while for Weierstrass they had been "the foundational principle" (Klein, 1926, p. 254). Pauli repeated his characterization in a letter to his PhD student Armin Thellung:[42]

> Since you were last in Zurich, I had some quite interesting correspondence with him [Dyson]. It seems to me, however, after this exchange that after all he has *not* done what is needed, because he focused too strongly on convergence proofs for power series (which is why I often jokingly refer to him as the "Weierstrass" of theoretical physics). For him it *remains* the fact

---

[41] Letter from Pauli to Dyson, February 18, 1951 (von Meyenn, 1996, p. 264).
[42] March 18, 1951 (von Meyenn, 1996, p. 276)).

that... *renormalization is only definable* through power series of the coupling constant... But it is now my impression that Dyson, with his method of power series (so far, of course this might change), is completely blocking his view (we need a "Riemann" and not a "Weierstrass" in the quantized field theory).

It should also be noted that Pauli was not accusing Dyson of lacking rigor, which would be the standard criticism of Dysonian perturbation theory today. In the historiography of mathematics, Weierstrass is considered to be the embodiment of rigor, while Riemann is viewed as a far more intuitive mathematician.[43] If anything Dyson was being too rigorous, or rather too formalistic, clinging to the reliable yet unintuitive method of series expansion, rather than seeking for physical insight. Dyson, however, was undeterred. In a letter dated April 15, 1951, to Hans Bethe (at Cornell, where Dyson had accepted a professorship for the fall), Dyson announced that he had completed his program of renormalizing the Heisenberg operators and then continued:

> The next problem is to prove the convergence of the perturbation theory expansion. I am convinced now that this can be done, and that the convergence will occur even for large values of the coupling constant. But the proof will be a long and very difficult piece of analysis...[44]

Dyson was not the only one trying to complete the mathematical foundations of renormalized, perturbative QFT and to identify the convergence of the series as the central missing element. Beginning in 1949, when Dyson had first shown the possibility of renormalizing the S-Matrix at each order, a similar line of research had been pursued at Cambridge, led by Nicholas Kemmer, from whom Dyson had originally learned QFT.[45] There was significant contact between Dyson and Kemmer's group,[46] and Dyson had even received an offer of employment from Cambridge, which he declined in favor of Birmingham (Schweber, 1994, p. 551). However, there was a slight difference in emphasis between Dyson's program and that of the Cambridge school. While Dyson had begun by extending his proof beyond the original specialization to the S-Matrix, the initial focus at Cambridge remained on the S-Matrix and the immediate application of the new renormalization methods to theories besides QED.

---

[43] See, e.g., (Birkhoff and Bennett, 1988, p. 147). Indeed, Weierstrass was known to point out gaps in Riemann's proofs (Cooke, 1984, p. 15).

[44] Cited in (Schweber, 1994, p. 565). Similar remarks are found in a letter to Abdus Salam, dated 14 May 1951, where Dyson also remarked that the convergence could not be proven by "some 'trick' method," as he had originally hoped (ASP, Dyson Correspondence).

[45] See Kemmer's Autobiographical Notes, p. 27, NKP.

[46] See (Kaiser, 2005, pp. 118–119) and (Close, 2011, pp. 63–64).

While Kemmer himself did not conduct active research on renormalized QFT,[47] he had, in January 1949, "2 1/2 men" studying the "Schwinger, Tomonaga etc. stuff" who "played with the obvious generalisations to meson theory."[48] Kemmer's PhD student, Paul Taunton Matthews (1949, 1950), investigated how widely Dyson's method of perturbative renormalization of the S-Matrix could be applied beyond QED, developing a categorization of quantum field theories into those whose S-Matrix could be renormalized and those that were nonrenormalizable (PhD 1950). Kemmer's group was joined in 1950 by Abdus Salam (PhD 1952), who closed a loophole in Dyson's proof of the renormalizability of the S-Matrix by showing how to deal with so-called overlapping divergences (Schweber, 1994, pp. 542–544). The question of overlapping divergences was also addressed, using a different method based on what would come to be known as the Ward identity, by John Clive Ward, who, while at Oxford, established strong ties to the Cambridge group (Ward, 2004, p. 9). In his paper, Ward (1951, p. 897) gave a succinct statement of the Cambridge program:

> The success of the concept of renormalization in giving a theory of spinor electrodynamics that is both finite and in good agreement with experiment has made the extension of the method to other field theory problems most desirable, particularly the extension to the problem of nuclear forces. The first step in this direction is to show that the renormalized S-matrix is finite.

Kemmer (whose group came to be called the "subtractionists" by other Cambridge physicists)[49] further cosupervised the PhD work of Richard Eden (PhD 1951), together with the most famous Cambridge physicist, Paul Dirac (Hamilton, 2009, chapter 10).[50] Eden stood in the tradition of Heisenberg's pure S-Matrix approach (Cushing, 1990, p. 69); his main focus lay on the analytic properties of the S-Matrix. He would later become an important figure in the further development of that approach. While it was not his primary concern, he prominently mentioned the convergence of the power series as an unsolved issue in a paper on the analyticity of the perturbative S-Matrix (Eden, 1952, p. 390). Finally, there was Jim Hamilton, who had joined the Cambridge faculty as a lecturer in early 1950. Hamilton had completed a PhD thesis in Manchester on Heitler's theory of radiation damping, a pre-renormalization attempt at

---

[47] In his autobiographical notes (p. 27), Kemmer complained that he was "failing to return to original research while performing [his] teaching duties", NKP.

[48] Letter from Kemmer to Peierls, January 31, 1949 (Lee, 2009, p. 166).

[49] Letter from Felix Pirani to Alfred Schild, November 12, ASchP, box 86-027/1.

[50] Dirac, as mentioned in the introduction, remained very skeptical towards renormalized QFT throughout his life.

removing the divergences in QFT scattering amplitudes.[51] At Cambridge, he joined Eden in studying the singularities and branching points of the S-Matrix. The question of the convergence of the power series he delegated to his first PhD student, the Australian Charles Angas Hurst (PhD 1952). Thus, in the spring of 1951, there were two people trying to prove the convergence of the Dyson series: Dyson himself in Birmingham and Hurst in Cambridge. And, almost simultaneously, both arrived at a negative conclusion.

2.4 Nonconvergence of the Perturbation Series

As fate would have it, Dyson spent the summer of 1951 in Zurich with his interlocutor and critic, Pauli. The Dysons were expecting a child in July. In April, Verena had moved to Zurich, where her parents lived. Freeman had accepted a professorship at Cornell, set to begin in the fall, so he wrapped up things in Birmingham and left for Zurich in June to join his wife (Dyson, 2018, p. 167). In the letter announcing his visit to Pauli, Dyson was still very optimistic regarding his program, while happily conceding that it might not lead to deep physical insights:

> As you know, my wife is in Zurich, and I am hoping to be able to leave England and join her some time before the end of June, with luck about June 20. I would be extremely happy if I could arrive while you are still in Zurich, and hear your views on the relation of mathematics to physics and on many other things.
> I have been meditating a little on the long-range problems of theoretical physics, especially the problem of understanding the renormalization program on a deeper level and escaping from the power-series expansions. It seems to me that the road ahead leads to a complete descriptive theory of elementary particles, based on the renormalization technique with power-series expansions. This theory may well be able to describe accurately the results of all possible experiments on the atomic scale, and still contain no explanation of the need for renormalization nor of the nature and variety of elementary particles nor of the mystic number 137. Beyond this descriptive theory, there is an absolute darkness.[52]

Esther Dyson was born on July 14, 1951, in Zurich. But her father also had time for frequent discussions with Pauli, "when Pauli would sometimes feel hungry in the middle of the afternoon and go out to an open-air café on one

---

[51] On which, see (Blum, 2017). Hamilton (1986) argues that damping theory actually contained premonitions of later dispersion-theoretical methods, thereby also explaining both why he continued working on damping theory for a few years, even after the development of renormalization theory, and why he was then predisposed to engage with dispersion-theoretical S-Matrix work.

[52] Letter of May 16, 1951 (von Meyenn, 1996, p. 305).

of the little streets near the Gloriastrasse for a dish of ice-cream."[53] These discussions over ice-cream finally led to an epiphany, as Dyson later recalled:

> In Zurich I discussed my ideas with Pauli, and Pauli remained sceptical. Pauli said he could not prove me wrong, but he had a strong feeling that the crucial... series that I assumed to be convergent would actually diverge. One afternoon in Zurich, while I was walking with Pauli, I suddenly saw that his intuition was right. I found a simple physical argument showing why the series could not converge. I explained the argument to Pauli, and he said, "I told you so."(Dyson, 2018, p. 168)

Rather than presenting a "long and very difficult piece of analysis," Dyson (1952) ultimately ended up submitting a two-page paper to the *Physical Review*, in which he presented his argument that the perturbation series does not converge, no matter how small the coupling constant.

The argument went roughly as follows: Take some physical quantity F, expressed as a power series in the square of the elementary charge e^2. If the perturbation series converges, then $F(e^2)$ is an analytic function of e^2, with a finite radius of convergence around the origin. Consequently, $F(-e^2)$ will also be a convergent power series. However, if indeed e^2 were negative, opposite charges would repel.

There would thus be, even in the vacuum state, a small, but finite, probability for the creation of a bunch of electron–positron pairs. Normally, with positive e^2, the electron and the positron in each pair attract one another – a deadly attraction, as they will immediately annihilate one another. However, with negative e^2, the electron and the positron would repel each other and fly off in opposite directions. They would then, in fact, tend to cluster with other particles of the same type – electrons with electrons, positrons with positrons. At some large number threshold, the electrostatic binding energy of these two clusters would exceed the rest mass energy needed to create all the pairs (since the binding energy grows exponentially with the number of particles), making this indeed a physically allowed transition.

For the creation of n pairs, such a transition would involve a factor of e^{2n} and would thus appear only at very high orders in perturbation theory; yet, these transitions could not be neglected, since they prevented the possibility of a stable ground state: once the two clusters were created, they would keep on growing quickly, since now the energy balance was positive even if one added just one particle at a time. The existence of nonnegligible contributions from

---

[53] This detail is from Dyson's unpublished Pauli Memorial Lecture "Our Stability is but Balance," which he gave in Zurich on February 18, 1974. A copy of the manuscript is held in the Albert Einstein Archives (call number 160–020). Apparently Dyson sent a copy to Helen Dukas, Einstein's former secretary and, after Einstein's death, the trustee of his estate.

very high orders of perturbation theory implied that the terms in the perturbation series start getting bigger again, once the order of perturbation exceeds the threshold for bulk electron–positron pair creation, in contradiction with the assumed convergence of the series.[54]

During the summer of 1951, Dyson also discussed his results with Abdus Salam, whom he had invited to visit him in Zurich.[55] After Salam returned to England, he reported to Kemmer:[56]

> I came back from the continent on the 19th... I saw Dyson fairly regularly I was all the time in Zürich spending almost 5/6 hrs. each day. He is now engaged on a most imp[or]t[an]t development. He is convinced (after repeated failures to prove the contrary) that the series expansions [sic] is never c[onver]g[en]t; at best it is asymptotic; This is the same result which Hurst has derived.... I saw Pauli for about half an hour...

Hurst must thus have obtained his result, which was then published in two papers (submitted on November 14 and 30, 1951), around the same time as Dyson. However, Hurst had obtained it in a very different, more quantitative manner. From Dyson's argument it was not clear how it could happen that the terms in the perturbation expansion would start increasing again; after all, the nth order of perturbation theory comes with a factor of g^n, and if the coupling constant g is less than one, this should lead to exponential decrease. However, as Hurst (1952b) could show, this decrease is more than counterbalanced by the increase in the number of Feynman diagrams, which goes more or less as $n!$. Of course one might hope that the contributions from the individual Feynman diagrams, when properly evaluated, might actually fall off faster than g^n. But Hurst (1952a) was able to show that this was not the case, at least for a toy model of a scalar field with a ϕ^3 self-interaction. The terms in the perturbation series would thus decrease only up to the order where the additional factor of g no longer sufficed to counterbalance the additional combinatorial factor of n. At order $n \approx 1/g$, the terms would start increasing again indefinitely, as in Dyson's heuristic argument.[57]

---

[54] Dyson's *reductio* proof of nonconvergence is sometimes misconstrued as taking this instability of the vacuum to be an inconsistency in and of itself, rather than merely being inconsistent with the assumed convergence of the series (Wightman, 1979, p. 994). This makes Dyson's argument seem weaker than it is, since one could actually imagine a consistent physical theory with an unstable vacuum.

[55] Letters from Dyson to Salam, May 14, and July 25, 1951, ASP, Dyson Correspondence.

[56] Letter of 22 August 1951, NKP, Salam Correspondence.

[57] Just like Hurst, Dyson had estimated that the perturbation terms would start increasing at order $1/g$ (order 137 for the case of QED). However, since he did not provide an argument for this value, it can be assumed that he simply took it from Hurst. This estimate is not yet mentioned in the previously quoted letter from Salam to Kemmer.

Dealing, as he did, only with a toy model, Hurst had hardly provided a rigorous proof of the nonconvergence of the power series in QED. Nevertheless, taken together, Dyson's and Hurst's results made it highly probable that the perturbation series in QED did indeed diverge. The program of proving the consistency of perturbative QED, pursued in Cambridge and by Dyson in Birmingham, had failed. What was one to make of this?

2.5 Asymptotic Series

Had this result been obtained 100 years earlier, it would have been viewed as utterly detrimental to the mathematical legitimacy of QED. In the early nineteenth century, a number of mathematicians, foremost among them Augustin Louis Cauchy, had pursued a program of placing 'on a firm, rigorous footing many mathematical methods and procedures that had up to then been used without sufficient theoretical justification" (Belhoste, 1991, p. 9). One of their primary objectives was to identify explicit criteria for determining whether an infinite series converges. The most important tool in this regard was the radius of convergence. Cauchy had shown, for example, that the Taylor series of a function $f(x)$ converges for real values of x smaller than the absolute value of the smallest complex number z for which $f(z)$ diverges.[58] We have seen the undiminished importance of the concept of radius of convergence in Dyson's argument for nonconvergence of the perturbation series.

The proponents of this "critical movement" sought to abolish the use of divergent series in mathematics. The Norwegian mathematician Niels Henrik Abel, another early champion of the new rigor, famously referred to divergent series as "the invention of the devil" (Kline, 1972, p. 973). Then, as now, solutions to differential equations were frequently constructed through a series expansion (Kline, 1972, section 21.6). Cauchy asserted that this method gave

> zero certainty that one has effectively solved the given equation, nor even whether that equation admits a solution at all.... The solution of differential equations by series was thus illusory, as long as one did not provide any means of ensuring that the series obtained are convergent, and that their sums are proper functions solving the given equations; so that it was necessary to either find such a means, or to look for a different method by which one could establish in a general manner the existence of proper functions solving the differential equations and calculate values that are arbitrarily close to those very functions. (Cauchy, 1840, pp. 327–328)

---

[58] This statement is sometimes referred to as the "Turin Theorem," as Cauchy was in voluntary exile in Turin at the time. A staunch royalist, he had left France after the revolution of 1830 (Belhoste, 1991, p. 152).

Cauchy was thus moved to give the first general existence theorems for solutions of differential equations, independent of the actual construction of such solutions.

The ban on divergent series was certainly also meant to extend to mathematical physics. Legend has it that "[a]fter a scientific meeting at which Cauchy presented his theory on the convergence of series, Laplace hastened home and remained there in seclusion until he had examined the series in his *Mécanique céleste*. Luckily every one was found to be convergent" (Kline, 1972, p. 972). This anecdote probably primarily refers to Taylor series, that is, expansions in an independent variable. However, Laplace (1798, book 2, chapter 5) had also used expansions in terms of a small constant parameter. He had studied the equations of motion for a body moving under the influence of both a primary force (such as the gravitational attraction of the sun) and a small perturbing force αQ, where α is a "very small constant coefficient, which, in the theory of celestial motion, is of the order of the perturbing forces" (1798, book 2, chapter 5, p. 235). The solutions were constructed iteratively and appeared as a power series (1798, book 2, chapter 5, pp. 242–243), which resembled the perturbation series of QED, down to the use of the letter α to denote the expansion parameter. One can thus easily imagine a nineteenth-century Dyson hastening home after Cauchy's presentation, only to find himself not quite as lucky as Laplace.

It was only in the late nineteenth century that this negative attitude started to relax again, as mathematicians came to accept that there are "divergent series that are useful in the representation and calculation of functions," and consequently formalized and developed the concept of an asymptotic series (Kline, 1972, p. 1103). One reason for this shift was that Cauchy's rigorous notion of convergence was not always helpful for actual calculations. In his *Méthodes nouvelles de la Mécanique céleste* ("New methods of celestial mechanics"), the French mathematician Henri Poincaré argued that mathematicians and natural scientists really had different notions of convergence:

> Between mathematicians and astronomers some misunderstanding exists with respect to the meaning of the term "convergence." Mathematicians who are mainly concerned with perfect rigorousness of the calculation and frequently are indifferent to the enormous length of some calculation which they consider useful, without actually thinking of ever performing it efficiently, stipulate that a series is convergent if the sum of the terms tends to a predetermined limit even if the first terms decrease very slowly. Conversely, astronomers are in the habit of saying that a series converges whenever the first twenty terms, for example, decrease rapidly even if the following terms might increase indefinitely.(Poincaré, 1993, p. 317)

In working with differential equations, "the infinite series which were used to find the solutions did not always yield useful approximations" (Schlissel, 1976, p. 309). A striking example is the three-body problem in Newtonian gravity, another famous "unsolvable" problem. In the early twentieth century, the Finnish mathematician Karl Sundman proved that an analytic solution exists: a convergent expansion in powers of $t^{1/3}$. However, it converged so ridiculously slowly that it was entirely useless for practical astronomy. Instead, astronomers continued to use mathematically divergent series (with t only appearing in the arguments of trigonometric functions), whose first terms gave very precise predictions for, for example, the lunar orbit (Barrow-Green, 2010). It was in this astronomical context that Poincaré first introduced the notion of asymptotic series (Kline, 1972, p.1104).

By the mid twentieth century, divergent series had become standard fare. Indeed, both Dyson and the Kemmer group were very likely familiar with the textbook *Divergent Series* by the Cambridge mathematician G.H. Hardy (1949), which asserts in its preface:

> Then came a time when it was found that something after all could be done about [divergent series]. This is now a matter of course, but in the early years of the century the subject, while in no way mystical or unrigorous, *was* regarded as sensational, and about the present title, now colourless, there hung an aroma of paradox and audacity.

Hardy's book begins with several examples of divergent series. And the example $f(x)$ he used to discuss the notion of asymptotic series (1949, p. 26ff) bears an uncanny resemblance to the diverging perturbation series:

$$f(x) = \sum_{n=0}^{\infty} n!(-x)^n \qquad (3)$$

Just like the Dyson series, this was a power series whose coefficients increased as (or rather, in this simple case, were equal to) $n!$. And Hardy showed that $f(x)$ was an asymptotic expansion of the function

$$f(x) = \int_0^{\infty} \frac{e^{-w}}{1 + xw} dw \qquad (4)$$

that is, the difference between the function and the first m terms of the series is less than the next term in the series,[59] which is equal to $(m + 1)!x^{m+1}$.

---

[59] This estimate holds only for x in the right half of the complex plane, i.e., for $x = re^{i\theta}$ with $|\theta| \leq \pi/2$, which of course includes the important case of x real and positive. For x in the left half of the complex plane, the upper bound on the remainder is weakened by a factor of $1/|\sin\theta|$. For real negative x, i.e., for $\theta = \pi$, the upper bound becomes infinite, and the series is no longer an asymptotic expansion of $f(x)$.

"There is therefore," Hardy wrote (1949, p. 28), "one sense at any rate in which the series 'represents' $f(x)$." He consequently used the same symbol, $f(x)$, for the series and the function.

The perturbation series thus had all the makings of an asymptotic series, and Hurst (1952a, p. 638f) concluded his second paper by observing:

> If it be granted that the perturbation expansion does not lead to a convergent series in the coupling constant... then a reconciliation is needed between this hypothesis and the excellent agreement found in electrodynamics between experimental results and low-order calculations. It is suggested that this agreement is due to the fact that the S-matrix expansion is to be interpreted as an *asymptotic* expansion in the fine-structure constant, so that the agreement mentioned is due to the fact that as the fine-structure constant is small, the error committed by stopping the calculations at any order of approximation is of smaller order than the last term calculated.

What did all of this mean for the consistency of QED? While neither Dyson nor Hurst actually provided a full proof of the divergence of the perturbation series, the agreement between these two very different arguments was nonetheless quite persuasive. And if one accepted their conclusion, one had to concede that the perturbation expansion could not be used to construct an exact solution of QED or prove the existence of such a solution.[60] However, the divergence of the perturbation series also did not imply that no such solution existed. If anything, the empirical success of the perturbation series suggested, as Hurst argued, that it *probably was* the asymptotic expansion of an exact solution.

This could be taken as a conclusion to the attempt at proving the consistency of QED: the existence of exact solutions had been shown to be likely; indeed, said existence could be considered – in a mathematically sophisticated way – the best explanation for the empirical success of QED. The divergence of the perturbation series thus did not really change the status of QED as a theory, nor did it have obvious implications for the applicability of QFT for the strong nuclear interactions. But it did significantly change the status of perturbation theory. Dyson had believed that QED was nothing more than the perturbative expansion and that consistency could and would be proved through perturbation theory. Now, perturbation theory had been demoted to a mere approximation method. And even though the results of Dyson and Hurst

---

[60] Under certain conditions one can reconstruct a function from its asymptotic series, but in the context of QFT such methods were only explored decades later. Using Borel summation of the perturbation series, one can reconstruct the known nonperturbative solutions from the perturbation series for some simple models in constructive QFT (Miller, 2017, p. 56). There is still some hope that generalized Borel summation techniques could be used to obtain exact solutions for more complicated and realistic models (Dunne and Ünsal, 2012, Section 8).

only addressed (and disproved) the mathematical convergence (in the sense of Poincaré) of the perturbation series, the expectations for its physical convergence immediately plummeted. For theories with strong coupling, "even the first approximation would be dubious" (Hurst, 1952a, p. 639).

The results of Dyson and Hurst thus emphasized that nonperturbative methods would be necessary – both to assess the consistency of QED and to extend QFT to the nuclear interactions. They even had some ideas what to look for: since exact solutions in QFT apparently did not have a convergent series expansion, even in the limit of a vanishing coupling constant, one should expect an essential singularity for e^2 or g^2 equal to zero.[61] According to Salam,[62] Dyson was "hopeful of great things" in this regard and "expect[ed] an essential singularity of the type $\exp(-1/e^2)$ at the origin." While the exponential of $1/x$ can be considered the standard example of an essential singularity, one can still appreciate Dyson's intuition: we will indeed encounter similar expressions in later sections, as we study the attempts to investigate consistency beyond the perturbation expansion.

But first we need to bid farewell to the protagonist of the first section. Dyson hardly participated in these investigations. In his short paper on the divergence of the perturbation series, he enthusiastically embraced the notion of open and closed theories and proposed that QED might not in fact be closed and that the equations of QED had no exact solutions. All that QED then did was give a useful approximation to an as yet undiscovered theory, for the construction of which "not merely new mathematical methods but a new physical theory is needed." Dyson had thus fully switched to Oppenheimer's view and stopped working on QED entirely (Schweber, 1994, p. 565). He did work on nuclear physics for a while longer (Dyson et al., 1954), and we will encounter him as a side character in later sections. But he ultimately moved on to other topics, removed from relativistic QFT, such as solid-state physics. As he wrote to Murray Gell-Mann in 1955:[63] "I have now left the hard territory of field theories and I am wandering happily in the dream-land of spin-wave interactions, where every Born approximation series converges..." In hindsight, Dyson would identify his discovery of the nonconvergence of the perturbation series as a decisive turning point – perhaps never more explicitly (and at a more suitable occasion) than in his 1974 Pauli Memorial Lecture in Zurich:[64]

---

[61] While Hardy's example, Equation (4), may at first glance appear to be perfectly well-behaved (and equal to 1) at the origin, things are not that simple: due to the branch cut on the negative real axis, the limit of $f(x)$ for $x \to 0$ is, in fact, not uniquely defined (Hardy, 1949, p. 27).
[62] Letter to Kemmer of August 22, 1951, NKP, Salam Correspondence.
[63] Letter of October 24, 1955, MGP, Box 6, Folder 33.
[64] Page 4, see footnote 53.

We entered in 1951 the modern era of particle physics, which resembles the modern era in art and music. Instead of trying to build a great coherent theory along classical lines, the physicists of the new era work with fragments of themes, with dissonant ideas placed in apparently random juxtaposition, just like modern composers.

It is thus to the modern era of particle physics, to the search for nonperturbative solutions, and back to Pauli that we now turn.

3 The Search for Nonperturbative Solutions

The apparent divergence of the perturbation series made perturbation theory irrelevant to the consistency question in QED. The remainder of this book will therefore focus on non-perturbative approaches to QED. "Nonperturbative" is used here in the widest possible sense, running the gamut from approximations still partially reliant on the perturbation series to exact results independent of any approximation at all.[65] This is a vast field – a fact that is easily overlooked when focusing solely on the dichotomy between lowbrow "conventional" QFT and highbrow algebraic QFT (Wallace, 2011; Fraser, 2011). Given this vastness, this chapter is probably the most impressionistic and idiosyncratic of the entire book. I will focus on those aspects of middlebrow nonperturbative approaches to QFT that arose in the early 1950s and were either directly connected to the question of consistency or will become relevant to it in later chapters. This will be something of a *tour d'horizon*, but it will start right where we left off – in Zurich, in 1951.

3.1 Källén and Thirring

While for Dyson the nonconvergence argument was "a terrible blow to all my hopes" (Schweber, 1994, p. 565), for Pauli the fun was just beginning. In a sense, it was history repeating itself. Some 20 years earlier, Oppenheimer had made his discovery that radiative corrections in QED lead to an infinite displacement in the spectral lines – as a postdoc with Pauli in Zurich.[66] Pauli had reacted by briefly turning Zurich into the center for studying the extent of these difficulties. The central papers establishing that Oppenheimer's result was no accident, but rather revealed a critical defect of QED, were written by postdocs

---

[65] Duncan (2012, p. iv) also remarks on the vagueness of the term "nonperturbative."
[66] The paper was submitted after Oppenheimer had returned to the US, but he later recalled that "when I came back I had practically everything that I was going to do on the self energy problem, the fields, but I didn't have it written up." Interview with Thomas S. Kuhn, November 20, 1963. AHQP, Microfilm 1419–04.

and visitors under Pauli's supervision in Zurich.[67] In 1951, Pauli reacted in much the same way to Dyson's result.

The physicists whom Pauli set to work were two young postdocs, Gunnar Källén from Sweden and Walter Thirring from Austria. Both arrived in Zurich to work with Pauli in late 1951,[68] and soon started investigating the status and the relevance of Dyson's heuristic proof of non convergence. It appears that Pauli had not been informed of Hurst's results. Salam had only seen Pauli for half an hour during his stay in Zurich, and Dyson recalled that he had tried to introduce the two, but Pauli had shown no interest.[69] Dyson himself, now a young father, did not keep in touch with Pauli – the next extant letter between the two after Dyson's departure from Zurich dates from December 1954. Consequently, when Thirring was assigned the task of proving Dyson's conjecture, he essentially repeated Hurst's analysis – using the exact same toy model. News of Hurst's work only reached Zurich when Hamilton sent Pauli a copy of Hurst's thesis.[70]

It was not entirely coincidental that Hurst and Thirring had used the same model – known for some time as the Hurst–Thirring Model (Bogoliubov and Shirkov, 1959, p. 352), but now usually referred to by the very matter-of-fact name of ϕ^3 theory. Ward (1950) had already shown that renormalization in ϕ^3 theory was far easier than in QED. The only perturbative UV divergence was the one associated with the simple second-order self-energy diagram. If one calculated that diagram and introduced a suitable mass counterterm, then the divergences in all higher-order diagrams were automatically removed: no higher-order corrections to the mass counterterm needed, and no charge renormalization at all. Thirring later introduced the term "super-renormalizable" for such theories, whose perturbative ultraviolet divergences can be removed in a finite number of steps. This feature of course made ϕ^3 theory ideally suited for studying the convergence of the perturbation series;[71] when estimating the higher-order terms relevant to addressing that question, one did not need to

---

[67] Ivar Waller (1930) demonstrated that Oppenheimer's result also applied to free electrons. Lev Landau and Rudolf Peierls showed that the divergences appeared in configuration space just as they did in occupation-number space (Landau and Peierls, 1930). Finally, Léon Rosenfeld found divergences analogous to those in QED for the gravitational coupling of photons (Rosenfeld, 1930) and for a nonrelativistic harmonic oscillator coupled to a quantized radiation field (Rosenfeld, 1931). Rosenfeld wrote this latter paper when he was already in Copenhagen but later recalled that it had been "directly inspired by Pauli," who wanted to see "how general is this divergence business" (Interview with Kuhn and Heilbron, see footnote 23).

[68] Letter from Pauli to Armin Thellung, 20 December 1951 (von Meyenn, 1996, p. 468).

[69] Interview by Jagdish Mehra, November 4, 1971, second side of tape, p. 29 of transcript. JMP, Box 99.

[70] Letter from Pauli to Källén, August 19, 1952 (von Meyenn, 1996, p. 708f).

[71] First attempts at a generalization to actual QED soon followed (Petermann, 1953)

concern oneself with the removal of new divergences. Hurst had thus ignored renormalization entirely in his argument, giving Thirring a reason to publish his work, which explicitly addressed this issue, after all.[72]

The bigger question – beyond more firmly establishing the divergence of the perturbation series – was whether there might exist solutions of QED that could not be expanded in a power series for a small coupling constant. This was the problem pursued by Källén.[73] He had been a postdoc with Pauli in 1950 and was now returning to work with him again for a few months.[74] He shared Pauli's predilection for mathematics (Jarlskog, 2014, p. 405ff) and now promptly took on the task of "looking behind the veil of the Dysonian power series." As the reader will have anticipated, Källén did not manage to prove or disprove the existence of exact solutions to the equations of QED. However, he did make substantial progress on the first major obstacle, formulating "a definition of renormalization that is not explicitly bound to these power series."[75]

A central problem in even contemplating the existence of nonperturbative solutions was the intimate connection between renormalization and perturbation theory. How would one even write down the field equations that one then hoped to solve nonperturbatively? One couldn't work with the unrenormalized Hamiltonian, because one did not know the value of the bare mass and charge. And one couldn't work with the renormalized Hamiltonian, because the renormalization constants were only given as perturbatively constructed power series. It should be noted that this conundrum was not directly connected to the fact that the renormalization constants were infinite in perturbation theory. What was needed was simply a nonperturbative method for determining the renormalization constants.

We will illustrate Källén's proposal (Källén, 1952) using the example of the electron wave function renormalization constant Z_2. Consider the renormalized field equations for the electron field, which contained the physical mass and charge (which one knew) as well as the renormalization constants (which one didn't):

---

[72] (Thirring, 1953). See also letter from Pauli to Källén, September 16, 1952 (von Meyenn, 1996, p. 730).

[73] In his paper, Thirring also made some vague remarks on the subject, invoking "preliminary investigations" that implied that the perturbation series diverges around *any* value of the coupling constant, not just around zero, i.e., for small coupling constants. "This would mean that if there are solutions at all they must be non analytic for all real values" of the coupling constant (Thirring, 1953, p. 35). But Thirring does not appear to have followed up on these investigations.

[74] Manifold biographical information on Källén can be found in (Jarlskog, 2014).

[75] Both quotes are from a letter dated December 23, 1951, from Pauli to his former assistant Markus Fierz, in which he reported on Källén's work (von Meyenn, 1996, p. 480).

$$\left(i\gamma^\mu \partial_\mu - m\right)\psi(x) = \left(e\gamma^\mu A_\mu(x) + \delta m\right)\psi(x) \tag{5}$$

Actually, these field equations do not contain Z_2, but Z_2 does appear in the renormalized equal-time anticommutation relations for ψ:

$$\{\psi_\alpha^\dagger(\mathbf{x},t), \psi_\beta(\mathbf{x}',t)\} = Z_2^{-1}\delta_{\alpha\beta}\delta(\mathbf{x}-\mathbf{x}') \tag{6}$$

Källén then formally integrated the field equation to obtain:

$$\psi(x) = Z_2^{-1}\psi^{\text{free}}(x) - \int S^{\text{ret}}(x-x')\left(e\gamma^\mu A_\mu(x') + \delta m\right)\psi(x')d^4x' \tag{7}$$

where $S^{\text{ret}}(x-x')$ is a retarded Green's function for the Dirac operator, and $\psi^{\text{free}}(x)$ is an operator that obeys the free field equations and the usual (unrenormalized) canonical anti-commutation relations.[76] This formal solution now explicitly contains Z_2, in order for ψ to fulfill the renormalized anticommutation relations (equation (6)). At this point Källén introduced an additional supplementary condition from which the renormalization constant could be determined:

$$\langle 0|\psi(x)|q\rangle = \langle 0|\psi^{\text{free}}(x)|q\rangle \tag{8}$$

This condition requires some unpacking. The states $|0\rangle$ and $|q\rangle$ are both eigenstates (vacuum and one-electron states, respectively) of the free Hamiltonian, again with the physical mass. The field $(\psi^{\text{free}})^\dagger$ thus turns $|0\rangle$ into an infinite superposition of states $|q\rangle$, and the right-hand side of Källén's condition is simply e^{-iqx} times an appropriately normalized spinor. The field $\psi^\dagger$ creates a far more complicated state when acting on $|0\rangle$, at least when expressed in terms of eigenstates of the free Hamiltonian. Källén's condition now stated that $\psi^\dagger|0\rangle$ was equal to $(\psi^{\text{free}})^\dagger|0\rangle$ *plus* additional terms that, however, contained no further (free) one-particle states $|q\rangle$.

We can now compare this with Equation (7) to see how Källén's condition determines the renormalization constant. When the operator in that equation acts on the vacuum, both the first and the second summands individually create more one-particle components than allowed by Equation (8). However, one could now choose the wave function renormalization constant Z_2 such that the one-particle components created by the second summand on the right-hand

---

[76] Note that $\psi^{\text{free}}(x)$ obeys the free field equations with the physical mass, not the bare mass, i.e., Equation (5) with the *entire* right-hand side set to zero. The physical picture is that in the free theory the mass does not need to be renormalized, and thus the mass appearing in the field equations is already the physical mass. One could, of course, also approach the free theory from the interacting theory. In that case, if one were to "turn off" the interaction, one would end up with a particle with the bare mass, but this would be a somewhat perverse way of smuggling divergences into the free theory. The free mass is thus always taken to be equal to the physical mass m.

side of Equation (7) are exactly compensated by the factor of Z_2^{-1} in the first summand.

Källén's renormalization conditions were inspired by the perturbative renormalization procedure; but it was by no means clear that there is a limit in which the two conditions are equivalent, and Källén did not expound on the connection between his conditions and perturbative renormalization in any detail. In 1956, four years later, Léon van Hove (on whom more in chapter 5) would still refer to the nonperturbative renormalization conditions merely as a "reasonable guess of Källén."[77] More importantly, the main question – whether or not the equations of motion, supplemented with the renormalization conditions, actually had nonperturbative solutions – remained entirely unanswered when Källén returned to Sweden on February 23, 1952.[78]

It was quite obvious which outcome Pauli was ultimately rooting for. At the International Physics Conference in Copenhagen in June 1952,[79] where both Källén and Thirring presented their results, Pauli made it clear that he "would not feel happy if it was proved that the renormalization procedure leads to a consistent description of electrodynamic phenomena without taking into consideration other elementary particles" (Kofoed-Hansen et al., 1952, p. 55). This statement sheds some further light on Pauli's motivation. In addition to rejecting the mathematical tool of the power series, Pauli was also unhappy with the picture of the world implied by Dyson's original program: a neatly compartmentalized world it was, with separate renormalized QFTs for the electromagnetic, nuclear, and possibly gravitational interactions, connected only by the plus signs in the Lagrangian. Pauli's vision had always been a holistic one.[80] Already when writing his foundational papers on QED with Heisenberg in 1929, he had complained not just about the theory's lacking consistency, but also about the fact that the Lagrangian consisted of "three independent summands, among which no logical relation, by which I mean an inner connection, is established."[81]

While Pauli's ambition was clear, in the summer of 1952 he and Källén were at an impasse – the existence of nonperturbative solutions in QED could neither be proven nor disproven. It would require some novel ideas to move things forward, and we will return to Pauli only in the second volume of this Element

---

[77] Notes on a conversation with van Hove taken by John Wheeler, Wheeler Relativity Notebook III, p. 254, entry dated February 15, 1956, JWP.
[78] Letter from Pauli to Thellung, February 29, 1952 (von Meyenn, 1996, pp. 565–566).
[79] The conference was part of the founding process of CERN.
[80] For more on how Pauli's desire to see QFT fail fits into his wider philosophical views at the time, see a forthcoming paper by Noah Stemeroff.
[81] Letter to Oskar Klein, February 18, 1929 (Hermann et al., 1979, p. 491).

In the meantime, others were changing the rules of the game and redefining what it would mean for QED to have exact solutions.

3.2 Green's Functions as Solutions

In 1952, Pauli and Källén were not the only ones looking for exact solutions. So was Abdus Salam. Salam had returned to Pakistan in the fall of 1951 to take on a professorship at his alma mater, the Government College at Lahore (Fraser, 2008, p. 104). Here he did his best to continue his research, but the College did not subscribe to research journals, and Salam was expected to manage the soccer team rather than work on quantum field theory (ibid., p. 106). Nonetheless, he managed to do some work after all, as evidenced by a letter to Kemmer, in which Salam gave updates on his progress in constructing solutions of QED.[82] He had heard of Källén's work, but had not seen the paper. Salam's approach was very different from Källén's, in that he continued to work with the perturbation series and tried to find explicit expressions for the remainder.

While Salam was initially quite optimistic, his attempts appear to have progressed no further than Källén's. His letter is, however, interesting for another reason: it differs from Källén not just in its relation to perturbation theory, but also in the underlying notion of solution. Rather then looking directly for solutions to the field equations (or the Schrödinger equation) of QED, Salam was attempting to construct Green's functions. This was becoming the dominant view of what it would mean to solve a QFT, as I shall discuss in the remainder of this chapter.

The new focus on Green's functions was hardly the first step in the demotion of the traditional dynamical equations of QFT. In the course of the 1930s and 40s, the central paradigmatic problem of QFT had shifted from solving the Schrödinger equation to constructing the S-Matrix, culminating in Dyson's 1949 renormalizability paper (Blum, 2017). When Dyson extended his renormalizability proof and methods beyond the S-Matrix, he had anticipated a return to normalcy in this regard. In order to treat bound-state problems, he had worked towards establishing a finite, renormalized Schrödinger equation, which could then be tackled using "all the well-known approximate methods, variation methods, iteration methods, and if necessary numerical integrations" (Schweber, 1994, section 9.16). This would have marked a revival of the tried and true methods of nonrelativistic quantum mechanics. But this revival did not materialize, as it proved to be far more efficient to adapt the methods

---

[82] Letter dated April 15, 1952, NKP.

developed for scattering and apply them to other problems, such as calculating the energies of bound states. This was a further reason for Dyson to abandon QED, as witnessed by a letter of January 14, 1952, from Peierls to R. H. Dalitz, another postdoc of his: "To my great surprise, [Dyson] seems to have lost interest in his programme for treating divergence, and partly he seems to feel that all that can be done with his technique could be done more simply with the Bethe–Salpeter equation" (Lee, 2009, p. 291). The Bethe–Salpeter equation is a striking example of techniques originally developed for relativistic scattering now being applied outside that domain. Indeed, Dyson (1949b) himself had unwittingly pioneered this sort of integral equation, so we need to revisit his paper in a little more detail.

It had been Dyson's central insight that Feynman's diagrammatic approach to perturbation theory could be used to decompose the terms in the perturbation series into several recurring factors. For example, one could identify in any Feynman diagram *vertex parts*, which consisted of an arbitrary number of vertices and internal lines but were connected to the rest of the diagram solely through two electron lines and one photon line – just like a single vertex. The factor Λ_μ corresponding to such a vertex part could be calculated independently of the rest of the diagram. One could then calculate that rest, with the vertex part replaced by an unmodified vertex γ_μ, and then afterwards insert for that unmodified vertex the pre-calculated quantity $\Gamma_\mu = \gamma_\mu + \Lambda_\mu$. Γ_μ might even include the sum of many different (or, in principle, all) vertex parts, thereby enabling the calculation of multiple perturbation terms in one go. A similar procedure could be used for electron (Σ) and photon (Π) self-energy parts, i.e., the parts of diagrams that were connected to the rest only through two electron or photon lines. Again, one could initially replace them by the free electron (S_F) and photon (D_F) propagators and then later insert the modified propagators S'_F and D'_F, defined by

$$S'_F = S_F + S_F \Sigma S_F \tag{9}$$
$$D'_F = D_F + D_F \Pi D_F \tag{10}$$

In this way, one could divide the construction of the S-Matrix into (a) the calculation of isolated vertex and self-energy parts and (b) the evaluation of "skeleton graphs," where all of the self-energy and vertex parts were replaced by free propagators and unmodified vertices, respectively. This was not just a calculational simplification: the self-energy and vertex parts were also where the divergences occurred. Isolating them allowed Dyson to apply renormalization methods without having to deal with arbitrarily complicated skeleton graphs.

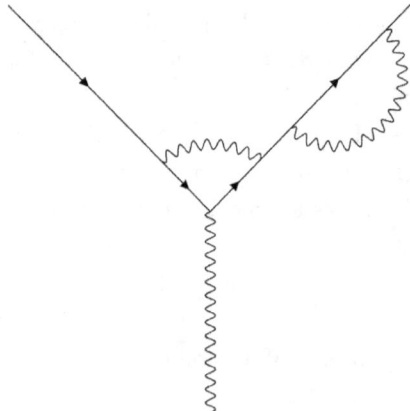

Figure 1 This (subpart of a) Feynman diagram has two electron and one photon external line and so could be considered a vertex part. However, due to the photon line that connects only to the right-hand external electron line, Dyson decomposed it into a proper vertex part and a self-energy part.

In order to make matters unambiguous, Dyson had to introduce the concept of a "proper vertex part." The simple definition of a vertex part given earlier also encompasses cases where one of the three lines joining the vertex part to the rest of the diagram contains a self-energy part (see Figure 1). Dyson instead considered such a structure to be the combination of a self-energy part and a proper vertext part, the latter being defined as a vertex part without self-energy insertions on its external lines. This made the distinction between what was a self-energy and what was a vertex part unambiguous. Dyson also introduced the concept of proper self-energy parts (Σ^* and Π^*), i.e., self-energy parts that could not be decomposed into two successive self-energy parts. An electron or photon line could, of course, contain arbitrarily many such proper self-energy parts, so that the defining equations for the modified propagators were adjusted to be:

$$S'_F = S_F + \sum_{n=1}^{\infty} S_F \left(\Sigma^* S_F \right)^n = S_F + S_F \Sigma S'_F \qquad (11)$$

$$D'_F = D_F + \sum_{n=1}^{\infty} D_F \left(\Sigma^* D_F \right)^n = D_F + D_F \Pi D'_F \qquad (12)$$

Finally, Dyson introduced the concept of "irreducible" vertex and self-energy parts.[83] These were (proper) vertex and self-energy parts that could not be obtained from simpler ones by insertion.

---

[83] This is not to be confused with the modern term "one-particle irreducible," which refers to a diagram that cannot be split into two separate diagrams by cutting a single line. Such diagrams would have merely been called "proper" by Dyson.

For the electron self-energy part, there was actually only one irreducible diagram, aside from the mass counterterm δm: the diagram where the electron emits and reabsorbs a virtual photon – two vertices, order e^2. This allowed Dyson to derive an explicit expression for Σ in terms of the modified propagators and the vertex function Γ:

$$\Sigma(p) = e_0^2 \left(\int d^4k \Gamma_\mu(p,p-k) S_F'(p-k) \gamma^\mu D_F'(k) \right) \tag{13}$$

For Dyson this was significant, because it provided a connection between results at lower orders of perturbation theory with results at higher orders. This connection could be used to make the induction step required to demonstrate that renormalization could be performed to an arbitrary order in perturbation theory. If one had the propagators and the vertex function up to order e^{2n}, Equation (13) gave the self-energy up to order e^{2n+2}. Inserting this into Equation (11) then gave the S_F' to order e^{2n+2}, i.e., one order (one factor of e^2) higher than the initial input.

Instead of evaluating Σ at some order of e^2, one could directly insert Equation (13) into Equation (11), obtaining not an explicit expression, but rather an integral equation for S_F':

$$S_F'(p) = S_F(p) + S_F \left[e_0^2 \left(\int d^4k \Gamma_\mu(p,p-k) S_F'(p-k) \gamma^\mu D_F'(k) \right) - \delta m \right] S_F'(p). \tag{14}$$

Here, the arguments of the vertex function Γ are the momenta of the two electron lines involved.[84] An analogous equation can be written down for the modified photon propagator:

$$D_F'(k) = D_F(k) + e_0^2 D_F(k) \text{Tr} \left(\int d^4p \Gamma_\mu(p,k-p) S_F'(k-p) \gamma^\mu S_F'(p) \right) D_F'(k) \tag{15}$$

Note that there is no such compact form for the equation governing the vertex function, since there is more than one (in fact, an infinity of) irreducible vertex parts. This is due to the possibility of the two electrons exchanging (multiply) crossed photon lines. In any case, while Dyson was aware of these relations, he never explicitly wrote them down, referring to them simply as "the integral equations." He saw no need to use or solve them in any way other than through the iterative construction of perturbation theory.

---

[84] It should also be noted that Dyson employed the Feynman gauge, where the photon propagator is proportional to $g_{\mu\nu}$, the Minkowski metric. Consequently, the photon propagator itself is not considered to carry Lorentz indices. Instead, one always contracts the Lorentz indices of the two vertices connected by a photon line.

Others, however, saw the appeal of actually working with these integral equations, as they could be used to extend the diagrammatic approach to non-scattering problems. Dyson had only considered integral equations for diagram parts that were more complex versions of the elementary constituents, the vertices and the propagators. However, one could also consider diagram parts that had no such elementary counterparts. In particular, an integral equation for diagrams (or parts of diagrams) with four external charged-particle lines could be used to determine the properties of two-particle bound states.[85] This is the previously mentioned Bethe–Salpeter equation, first derived in early 1951 by Hans Bethe and E.E. Salpeter at Cornell on a purely diagrammatic basis (Bethe and Salpeter, 1951; Salpeter and Bethe, 1951).

Just as with the vertex, there was an infinity of irreducible diagrams with four external lines, due to the possibility of exchanging crossed photons.[86] Pioneering the use of approximation methods for constructing (as opposed to just for solving) Dysonian integral equations, Bethe and Salpeter considered only one irreducible diagram: the simple one-photon exchange. The only reducible diagrams one could obtain from this were those where the photon lines exchanged between the two electrons did not cross (Figure 2), which Salpeter and Bethe (1951, p. 1234) aptly named "ladder-type graphs." By considering only this class of diagrams, they were able to obtain an explicit, closed-form integral equation as a starting point for their treatment of bound states.

As long as these integral equations concerned only the calculation of (parts of) Feynman diagrams, it was not at all clear what more fundamental role they might play within the structure of QFT. This changed after Salpeter presented his work with Bethe at the Institute for Advanced Study:

> Then finally I went and gave a formal colloquium at the Princeton Institute for Advanced Studies. What happened is somewhat like the apocryphal stories you hear about the Institute at the time – where they interrupt the speaker and don't let him finish and start writing the next paper right there and then. And that's almost what happened to me. I gave a colloquium on the equation,

---

[85] The four external lines correspond to the two particles (incoming and outgoing) that are supposed to form the bound state. I (and Bethe and Salpeter) thus speak of charged particles more generally, rather than specifically of electrons, since two electrons (or an electron and a positron) cannot form a stable bound state, as opposed to, say, an electron and a proton.

[86] It should be noted that the meaning of the term "irreducible" here deviates even more strongly from its current usage. For Bethe and Salpeter, "irreducible" diagrams were those diagrams that could not be obtained from simpler ones by inserting modified vertices, propagators or four-point diagram parts. Equivalently, this meant diagrams that could not be "split into two simpler graphs by drawing a line which cuts no quantum [photon] lines at all and *each of the two* particle lines only once" (Salpeter and Bethe, 1951, p. 1232, emphasis added). This (by modern standards) unusual use of "irreducibility" is also pointed out explicitly in the editor's remark of (Salpeter, 2008).

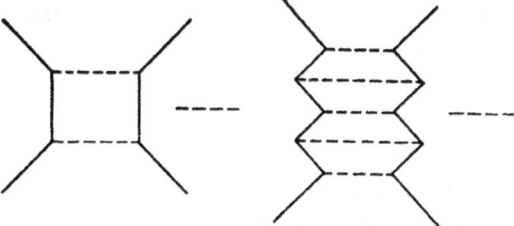

Figure 2 Some exemplary ladder diagrams, the class of diagrams that Salpeter and Bethe (1951, p. 1234) considered for setting up their integral equations. The essential point is that the ladders on the rung never cross. Note that the exchange particles are represented by dashed lines – the Bethe–Salpeter equation was supposed to cover both the exchange of photons and of (scalar) mesons. It is not clear what the dashed lines between the diagrams are supposed to represent.

and both Francis Low, who was already fairly well-known at the time, and Murray Gell-Mann, who was just some bright youngster from somewhere or other at the Institute, were in the audience, and got incensed enough at this sort of intuitive plausibility argument that they almost already started making their rigorous field-theoretic derivation of it, right there on the blackboard.[87]

Gell-Mann and Low's field-theoretic derivation of the Bethe–Salpeter equation was based on the realization that functions such as S'_F or the Bethe-Salpeter four-point function could not just be defined diagrammatically but also explicitly and compactly, without any reference to a perturbative expansion, as matrix elements of the Heisenberg-picture field operators. Their reconstruction was based on what is now known as the Gell-Mann-Low theorem, which relates the ground state of the free Hamiltonian (the interaction-picture vacuum) to the ground state of the full, interacting Hamiltonian (the Heisenberg-picture vacuum). Recall that Källén had worked *only* with the vacuum of the free Hamiltonian. His rationale had been that, when calculating S-matrix elements, one anyway assumes that the interaction vanishes adiabatically (i.e., infinitely slowly) for $t \to \pm\infty$. In this limit, the free and interacting Hamiltonians thus coincide, and so do their vacua. Since the states do not evolve in time in the Heisenberg picture, one can simply take the two vacua to be equal. Gell-Mann and Low (1951) acknowledged this approach but felt it was inadequate for the treatment of bound states, where the interaction could not be taken to vanish asymptotically. Using the expression they obtained for the relation between the two vacua, they were able to, e.g., identify the modified electron propagator (in

---

[87] Interview with Salpeter by Spencer Weart, www.aip.org/history-programs/niels-bohr-library/oral-histories/4854.

configuration space) with the time-ordered vacuum expectation value (VEV) of Heisenberg field operators:

$$S'_F(x,x') = \langle 0|T\left(\psi_0(x)\overline{\psi_0}(x')\right)|0\rangle \tag{16}$$

This expression was perfectly analogous to the free field theory: if one takes the operators and states on the right-hand side to be those of the free theory, then one has the free propagator S_F on the lefthand side, instead of S'_F.[88]

It was then Julian Schwinger at Harvard, who – in parallel with the work of Bethe, Salpeter, Gell-Mann, and Low – developed a complete reformulation of QFT based on such time-ordered VEVs and the equations determining them. He also introduced the custom of referring to these VEVs as Green's functions,[89] although, strictly speaking, that designation only applies to free fields, and even then only to the two-point functions, the propagators.[90] Still, I will adopt Schwinger's terminology in the following.

Schwinger (1951a) derived equations for the Green's functions directly from the Heisenberg field equations, that is, without taking a detour via the perturbative calculation of the S-Matrix. His derivation is somewhat obscured by his reliance on his new formulation of quantum theory based on the "quantum action principle."[91] But the general idea can be extracted from these idiosyncrasies, as I will do in the following. The most straightforward way to go from the field equations to equations for Green's functions is to take the VEV of the field equations. This just gives, however, an equation for the VEV of a single field operator, which is usually just equal to zero. Schwinger thus introduced an artifice into the field equations – namely, inhomogeneous terms interpreted as external (nondynamical) sources.[92] For example, the field equation for the electron now read:

---

[88] Note that this equation uses the convention – introduced by Gian Carlo Wick (1950, eq. 7'), but not yet in common use by 1951 – that the time ordering operator, when applied to fermions, gives a minus sign for each change in the order of the fields.

[89] That this somewhat idiosyncratic extension of the notion of a Green's function goes back to Schwinger is confirmed by Nambu (1955, p. 396). Indeed, Schwinger (1996) appears to have developed a certain fondness for George Green, considering himself "largely self-educated" just like the "miller of Nottingham."

[90] The fact that the free propagators are Green's functions was first highlighted by Feynman (1949, p. 750).

[91] (Schwinger, 1951c). The quantum action principle was Schwinger's attempt to translate Feynman's path integral approach "into a rigorous and conventional language" (letter from Dyson to Peierls, September 23, 1950 (Lee, 2009, p. 249)). While Schwinger's approach may not be that well-known today, many of its elements can be found in contemporary path-integral-based presentations of QFT.

[92] The reader may well wonder why Schwinger did not simply multiply both sides of the field equations by another field operator before taking the VEV, in order to obtain higher Green's functions. Formally, this would yield the same Schwinger-Dyson equations, but ambiguities

$$\left[\gamma_\mu \left(i\partial^\mu - e_0 A_0^\mu\right) - m_0\right] \psi_0(x) = \eta(x) \tag{17}$$

with a spinorial source term η. An analogous external vectorial current J_μ was included in the field equation for the electromagnetic field. With external sources, the VEV of a single field operator was no longer identically zero; instead, it became a functional of the external source. Schwinger could now show that functional derivatives with respect to the external sources increase the number of field operators in the VEV.[93] So, for example,

$$-i\frac{\delta}{\delta\eta(x')}\langle 0|\psi_0(x)|0\rangle = \langle 0|T\left(\psi_0(x)\overline{\psi_0}(x')\right)|0\rangle = S'_F(x,x') \tag{18}$$

after the external spinorial source η is set to zero. In this manner, one could take the VEV of the field equations and then apply functional derivatives to get equations for Green's functions involving more fields. A Green's function involving n fields would come to be known as an n-point function, making, e.g., the propagators two-point functions. The equation for the electron propagator S'_F read:[94]

$$\left[\gamma_\mu \left(i\partial_\mu - e_0 A_0^\mu\right) - m_0 + ie_0\left(\delta/\delta J_\mu(x)\right)\right] S'_F(x,x') = \delta(x-x') \tag{19}$$

By applying multiple functional derivatives one could, in principle, construct equations for Green's functions involving an arbitrary number of fields, though Schwinger stopped at the equation for the electron four-point function, which he dubbed the "'two-particle' Green's function."

How were Schwinger's functional differential equations related to Dyson's integral equations? Schwinger (1951b) was able to establish the connection by replacing the functional derivatives of the electron propagator with respect to the external current[95] by an integral over Dyson's vertex function Γ. This could be read as a definition of the vertex function within Schwinger's framework:

---

arise from multiplying operators at the same space-time point. Schwinger was well aware of this difficulty. Several months earlier, he had completed a paper on the definition of the four-current, which centrally involves the product of two electron operators. We will return to this issue in chapter 4 of the second volume.

[93] This was the step for which Schwinger had to invoke his action principle. Nowadays, the effect of the functional differentiation is usually determined from a generating functional, an approach first introduced by Symanzik (1954).

[94] One can see here, with the delta function on the righthand side, why Schwinger chose to refer to S'_F as a Green's function. Only the functional derivative (as opposed to another partial derivative) on the left-hand side spoils the identification of S'_F as a Green's function in the traditional sense.

[95] As it turned out, Schwinger's equations did not actually contain functional derivatives with respect to the spinorial sources η. The redundancy of these spinorial sources was soon pointed out by Utiyama et al. (1952).

$$\left(\frac{\delta}{\delta J_\mu(z)}\right) S'_F(x,y) = e_0 \int D'_F(z,z') S'_F(x,x') \Gamma^\mu(x',y',z') S'_F(y',y) d^4x' d^4y' d^4z' \tag{20}$$

The connection between the equation for the two-particle Green's function and the Bethe–Salpeter equation could be established in a similar manner (Karplus and Klein, 1952). Dyson's and Schwinger's equations were now viewed as the integral k-space and (functional) differential x-space versions, respectively, of what came to be known as the Schwinger–Dyson equations. This set of equations, in principle, determined the vacuum expectation values of arbitrary combinations of field operators. And soon such a full set of Green's functions came to be viewed as a satisfactory and complete solution of a QFT. The first written endorsement of this view that I could find is the aforementioned letter from Salam to Kemmer:

> *Any* matrix element for a scattering process or even bound state can be extracted from the general kernel [$\langle 0|T\phi(x)\phi(y)\cdots\phi(z)|0\rangle$] Thus if we can show that renormalization works for a general [$\langle 0|T\phi(x)\cdots\phi(z)|0\rangle$] we are home.

This view of QFT gained considerable traction over the next few years, culminating in a formal proof by Arthur Wightman (1956) that, at least for a neutral scalar field, one could indeed get a solution to the field equations from a full set of n-point functions.[96] This a result came to be known as the reconstruction theorem.

3.3 Solving the Schwinger–Dyson Equations

The Schwinger–Dyson equations provided a new, nonperturbative formulation of renormalized QFT based on time-ordered VEVs of field operators, which rivaled Källén's more traditional formulation (based on the field equations) and largely eclipsed it, not least due to Schwinger's many disciples: Paul Martin, who obtained his PhD with Schwinger in 1954, recalls that in the early 1950s, QFT at Harvard was a "religion" with "its own golden rule – the action principle – and its own cryptic testament – On the Green's Functions of Quantized Fields" (Martin, 1979, p. 70). The status of Schwinger's and Källén's approaches in 1950s field theory is perhaps best encapsulated in the mock template for a generic field theory paper that appeared in the *Journal of Jocular*

---

[96] Wightman's functions were the vacuum expectation values of products (not time-ordered products) of field operators and thus not exactly the same as the Green's functions discussed in this section. On the relationship between the Green's function and the Wightman function approach, see (Wightman, 1976, pp. 213–222).

Standard form FT/3.

Title: .... in field theory.
Author: ....

According to Schwinger

(1)

whence

(2)

hence

(3)

Thus[1]

(4)

which is not inconsistent with the assumption that

(5)

In virtue hereof

(6)

whence

(7)

hence

(8)

---

[1] Cf. also Källén, Ark. f. astr., mat. och fys. ....

Figure 3 Snippet from page 12 of the *Journal of Jocular Physics* of October 7, 1955. A copy is available digitally at the CERN archives. Reproduced courtesy of the Niels Bohr Archive, Copenhagen.

Physics, a spoof publication produced by a number of physicists in Copenhagen on the occasion of Niels Bohr's 70th birthday. The paper opens with a reference to Schwinger, while a later footnote refers to a paper by Källén in the Swedish journal *Arkiv för Matematik, Astronomi, och Fysik*, whose relative obscurity is highlighted by the incorrect ordering of the names of the disciplines in the citation (Figure 3).

Also with regard to renormalization, the Schwinger–Dyson equations were easier to use than Källén's field equations – renormalization proceeded (just as in Dyson's original formulation, in fact) in terms of propagators and vertex functions rather than of in terms of fields:[97]

$$S'_F \to Z_2 S_{FR}$$
$$D'_F \to Z_3 D_{FR}$$
$$\Gamma^\mu \to Z_2^{-1} \Gamma^\mu_R$$

where the index R indicates a renormalized quantity.

But did the establishment of the Schwinger–Dyson equations bring physicists any closer to finding exact solutions of QED? The immediate answer is: no, to the contrary. While the Heisenberg field equations could at least be written in closed form, the Schwinger–Dyson equations were an infinite set of coupled integral or functional-differential equations, with no obvious recursion relations among them. Effectively, this meant that there was really no realistic hope of finding an explicit, exact solution. In a sense, this meant the end of investigations into the consistency of QED. No one really ever tried again to construct exact solutions starting from the field equations, like Källén had, nor to somehow construct a full set of Schwinger–Dyson equations of QED. In another sense, however, the debate on consistency had only just started – after all, there is quite a bit of book left.

In the following chapters, we will encounter a variety of reactions to this new situation, with the perturbation series apparently divergent and other exact solutions out of reach. It should not be forgotten that the main reaction was probably the one that Walter Thirring (2008, p. 100) in his autobiography attributed to Feynman: "I don't give a shit whether it converges!" Even among those physicists who felt that probing the consistency of QED was still a worthwhile endeavor, there were differing views on how to go about it. One approach was to study approximate solutions other than the perturbation series. In the second volume of this Element we will see how such nonperturbative approximate solutions related to consistency questions. But it was anyway clear that new approximation methods would be needed, if one wanted to describe the nuclear interactions using quantum field theory. And it was here, as a starting point for new approximation methods, that the Schwinger–Dyson equations truly showed their usefulness.

---

[97] Note that the equations for the renormalized Green's functions (as opposed to the field equations) tend to contain extraneous renormalization constants. As a result, the Schwinger–Dyson equations were still frequently written in terms of the unrenormalized functions.

We have already mentioned the ladder approximation that Bethe and Salpeter applied to the equation for the four-point function, providing a starting point for the treatment of bound states in relativistic QFT. However, approximations could also be used to calculate Dyson's vertex function and propagators – and thus to redo the calculation of the S-Matrix with new methods that could be compared to, and ideally went beyond, perturbation theory. The first major attempt to use the Schwinger–Dyson equations to actually calculate the S-Matrix was undertaken by Samuel Edwards (1953), another member of the Kemmer group in Cambridge, where he was a PhD student under the supervision of Jim Hamilton. For the academic year 1951/52, he was awarded a J.H. Choate fellowship for students from Cambridge to spend a year at Harvard, from where he wrote a letter to Kemmer:[98]

> On arrival at Harvard Schwinger [...] said I must do field theory, and set me the task of calculating the proton-neutron mass difference. The earlier approaches to this problem [...] are by subtraction, but my method is to use the integral equations for Γ_μ (implicit in Dyson and in Schwinger's paper in the Proc. Nat. Acad. Sci.) [...] This equation I believe is insoluble (i.e., has no solution). But there are interesting considerations in hand which may get a solution. [...] I do not know of any other attempt to solve the field equations *as they stand*, so it seems to be breaking new ground.

The new ground eventually broken by Edwards essentially consisted in transferring the ladder approximation to the calculation of the vertex function Γ.[99] The reader will recall that there are infinitely many irreducible vertex parts, just as there are infinitely many diagram parts with four external electron legs, due to the presence of crossed photon lines. In full analogy to Bethe and Salpeter, Edwards retained only the simplest irreducible vertex part, where a single photon joins the two electron lines. The corresponding reducible diagrams are once again those with a ladder structure of nonintersecting photons exchanged between the two electron lines. Using this approximation, Edwards obtained a compact Schwinger–Dyson equation for the vertex function. The derivation of this equation is straightforward and closely mirrors Dyson's derivation of the equations for the propagators: calculate the modified vertex function in the ladder approximation; add an additional photon line connecting the two electron lines; get back the same vertex function, since the resulting diagrams once again

---

[98] Letter from February 2, 1952, NKP.
[99] I will describe Edwards' approach using conventional Feynman–Dyson diagrammatic language, even though he presented it in the rather different Schwinger style. As he joked in the margins of the letter to Kemmer when using Schwinger's notation G for the electron propagator: "Dyson's S_F (but now I work with Schwinger!)".

will have no intersecting photon lines and were thus fully accounted for in the original calculation. The integral equation obtained by Edwards then reads:[100]

$$\Gamma_R^\mu(p,q) = \gamma^\mu Z_2^{-1} - ie^2 \int \Gamma_{\nu R}(p, p-l) S_{FR}(p-l)$$
$$\times \Gamma_R^\mu(p-l, q-l) S_{FR}(q-l) \Gamma_R^\nu(q-l, q) D_{FR}(l) d^4l \quad (21)$$

While this equation is nice and compact, it *is* a nonlinear integral equation, further coupled with the equations for the propagators. Nevertheless, it was now the most promising starting point for obtaining approximate solutions of QED, and we will be encountering several such attempts in this book. Edwards's approximation was quite drastic: replace all of the Green's functions in the integrand with the unperturbed expressions, except for one of the vertex functions. This resulted in a linear integral equation for the vertex function, which did yield solutions in limiting cases such as $p = 0$. Edwards's solutions could reproduce the results of perturbation theory in that same limit when expanded in a power series, but he also observed hints of nonanalytic behavior. While Edwards's approximation still relied on perturbation theory for its formulation – the terms (or diagrams) he included in his calculation could only be identified within a perturbative expansion – it did represent a "partial departure from the expansion method" (Edwards, 1953, p. 285).

Yet, as Edwards remarked in his letter to Kemmer, exact solutions of QED remained as remote as ever. While Edwards's approach went beyond perturbation theory, there was no clear path for iterating his approximation to obtain an exact solution. Dyson's approach in perturbation theory – proving the convergence of an approximation method to demonstrate the existence of solutions – could not be pursued in the context of Green's functions and the Schwinger-Dyson equations. In the early 1950s, the search for exact solutions to QED ended. Exact solutions were now at best to be hoped for in less complicated toy models, as we shall see in the second volume of this Element. If one wanted to tackle the consistency of QED directly, one had to move from the ultimately intractable level of the equations to the theory's axiomatic foundation, as discussed in chapter 5.

But first, let us take stock: after the success of renormalized QED it had seemed natural for Dyson and the Kemmer group to investigate whether the theory was now indeed fully consistent. That effort had failed, and any further attempt to prove the consistency of QED through the existence of solutions

---

[100] Edwards wrote his equation for the *renormalized* Green's function, which brings in an additional factor of Z_2^{-1} in the first summand on the right-hand side. Note that we are using a different convention than Edwards: in our notation, the arguments of the vertex function are the two electron momenta; in Edwards's notation, it is one photon and one electron momentum.

appeared to be doomed to failure. So far, I have argued that the question of the consistency of QED (at least at the level of the dynamical equations) had become intractable.

At the same time, one can also argue that the question of the consistency of QED had now become physically irrelevant. Recall that before renormalization, there had been a clear physical interpretation of the putative inconsistency of QED; namely that the theory was unable to treat radiation reaction. With the calculation of the Lamb shift, however, QED had shown itself to be splendidly able to do just that. There still appeared to be something that perturbation theory could not capture, but it was entirely unclear what physical effects were supposed to be encoded in the nonanalytic or essentially singular parts of the Green's functions.[101] QED was, of course, empirically incomplete (or "open", to use Oppenheimer's terminology), but it seemed highly implausible that nuclear physics was hidden in the nonperturbative nooks and crannies of electrodynamics. The consistency of QED had not been proven – but was there still reason to doubt it?

There was. While much of the work on Green's functions discussed in this chapter was not (primarily) concerned with consistency issues, there were still those that cared. Källén had never given up on proving the inconsistency of QED. And the possible relevance of the consistency question for a theory of the strong interaction was still undeniable; this aspect would draw other physicists, most notably Gell-Mann and Low, to continue in its investigation. For those physicists still interested in the consistency of QED, the attention now shifted from the existence of solutions to what was arguably still the theory's most controversial aspect: infinite renormalization. If QED was to run into manifest difficulties of consistency, it would be in the context of renormalization, or in the domain of high energies where the divergences that renormalization was supposed to cure seemed to originate. We shall discuss this new focus of the inconsistency debate in the following chapter, before turning to the other directions mentioned earlier.

4 Infinite Renormalization and UV Behavior

Källén had neither been able to find exact solutions of QED, nor to prove their nonexistence. But he did not give up on probing the consistency of QED. In 1953, he turned his attention to what was widely considered the greatest weakness of QED: infinite renormalization. As mentioned in the introduction, for

---

[101] It was more than two decades later that matrix elements with an essential singularity for $\alpha = 0$ would come to be associated with tunneling transitions between topologically distinct ground states ('t Hooft, 1976).

some physicists, like Dirac, the entire procedure of infinite renormalization was so suspect that they rejected renormalized QFT altogether. For those like Källén, who actually worked with QFT, things were not as clear-cut. At least mathematically, the handling of infinities in perturbative QED seemed well under control through the "formalistic" regularization techniques of Pauli and Villars (1949).

Consequently, when Källén attempted to show that the infinities involved in renormalizing QED were not merely an artifact of the perturbation approximation but actually persisted even in his nonperturbative scheme, he did not view this as an inconsistency proof in itself. But it did address the most obvious weak spot of QED. For Källén, as we shall see in later chapters, this represented a first step, a lemma of sorts, in obtaining a full inconsistency proof. Before we look at Källén's investigation of the renormalization constants, we need to first look at the novel methods he used in this investigation.

4.1 Spectral Representations and the Källén-Lehmann Representation

More than just a new approach, investigating the finiteness of the renormalization procedure was a definite step away from a critical investigation of the solvability of QED. In Källén's nonperturbative formulation, the renormalization condition could only be formulated under the assumption that there were exact energy eigenstates in QED – specifically, the vacuum and physical one-particle states. While not much needed to be assumed about these states – not even, as we shall see later, that they could be normalized to one – their existence could no longer be called into question. Indeed, Källén's attempt to obtain a non-perturbative estimate for the renormalization constants rested on an even stronger assumption; namely, that there existed a complete set of states $|n\rangle$, each corresponding to a specific eigenvalue of energy-momentum p_n^μ.

With this assumption, Källén (1952) was able to derive an expression for the commutator of two Heisenberg field operators in terms of matrix elements of these operators for the states $|n\rangle$. Since this expression thus involved the spectrum of the Hamiltonian, it came to be known as the spectral representation. The spectral representation was later re-derived[102] in a more straightforward manner and in a simpler context (not QED, but scalar field theory) by Harry Lehmann (1954), on whom more later. Since Lehmann's treatment is significantly more transparent (justifying the later designation as

---

[102] On the context of Lehmann's derivation, see (Blum, 2019).

the Källén–Lehmann representation),[103] and does not differ in substance from Källén's original treatment, I will rely on Lehmann's presentation in discussing the spectral representation. The spectral representation for the VEV of the commutator of scalar Heisenberg fields ϕ is derived by inserting a complete set of energy-momentum eigenstates between the two operators, giving

$$\langle 0 |[\phi(x), \phi(x')]| 0 \rangle = \int_0^\infty \Delta(x,x'; \kappa^2)\rho(\kappa^2)d(\kappa^2) \tag{22}$$

where $\Delta(x,x'; \kappa^2)$ is the commutator of a *free* scalar field with mass κ^2, and the spectral function ρ is an unknown function determined by the actual spectrum of energy-momentum eigenstates.[104] The spectral function is implicitly defined by the relation

$$\int_0^\infty \rho(\kappa^2)\delta(k^2 - \kappa^2)d\kappa^2 = (2\pi)^3 \sum_{p_n=k} \langle 0|\phi(x)|n\rangle\langle n|\phi(x')|0\rangle e^{ik(x'-x)} \tag{23}$$

where the sum is over all states n belonging to some four-momentum eigenvalue k. Despite appearances, the right-hand side is also independent of x and x', since the exponential precisely cancels the space-time dependence of the matrix elements. From this equation, one obtains the value of $\rho(\kappa^2)$ at the

---

[103] The list of physicists that discovered the spectral representation can be extended even further. In Japan, two students of Shoichi Sakata, Hiroomi Umezawa and Susumu Kamefuchi, used a spectral representation to prove the theorem – discussed in more detail later – that the renormalized charge in QED is always smaller than the bare charge (Umezawa and Kamefuchi, 1951). Källén, who described himself as "rather particular about 'priority problems'" (Letter to Arthur Wightman, November 28, 1960, GKA), later freely acknowledged the priority of Umezawa and Kamefuchi in deriving the spectral representation. Kamefuchi also recalls that Wightman later praised their paper as "the beginning of axiomatic field theory" (Fax from Kamefuchi to the author, May 24, 2022). The spectral representation also appears in the 1954 paper by Gell-Mann and Low, discussed later. Gell-Mann later claimed that they developed the spectral representation independently of Källén (Interview with Gell-Mann by Sara Lippincott, July 17/18, 1997, p. 23, Caltech Oral Histories, https://oralhistories.library.caltech.edu/228/). This seems plausible, given that, as early as 1951, Gell-Mann and Low were making fruitful use of spectral decompositions in their field-theoretical derivation of the Bethe–Salpeter equation (Gell-Mann and Low, 1951, eq. 25). However, since they were there dealing with four-point functions, which cannot be rewritten as an integral over free functions with different rest masses, they did not in that paper obtain what we would now consider a spectral representation. Finally, in a letter to Arthur Wightman, Sam Schweber lists Wightman alongside Källén, Lehmann and "Gell-Mann and Low II" as having "told us the general form for these functions" (AWP). The letter is undated but contains remarks on the proofreading of Schweber's QFT textbook. In that book (Schweber et al., 1955, p. 386), Schweber also credits Wightman with having derived the spectral representation, referencing an unpublished 1952 manuscript, which is, however, not extant.

[104] Also, the reader should note again that these are renormalized field operators. There is a different convention, used by Schwinger, where the spectral function is defined through the unrenormalized operators. See the 1956 lecture notes *Differential Equations of Quantum Field Theory – A Set of Lectures Given at Stanford University*, Section VIII, JSP. I would like to thank Porter Williams for making these lecture notes available to me.

point $\kappa^2 = k^2$. To get the full spectral decomposition, one must thus apply this equation for all possible four-momentum eigenvalues to get the full spectral decomposition. In any case, ρ is a sum of absolute values and therefore positive definite.

When taking the limit of equal times, $x_0 = x_0'$ in equation 22, the interacting commutator on the left-hand side and the free commutator Δ on the right-hand side are both zero; taking this limit instead for the time derivative (with respect to x_0, say) of both sides, one gets the equal-time canonical commutator for the renormalized fields. This is a known, non-vanishing expression (proportional to $\delta^{(3)}(\mathbf{x}-\mathbf{x}')$). In fact, the expression is nearly identical to the free commutator on the right-hand side – it only differs by a factor of the field renormalization constant Z_3. Källén and Lehmann thus obtained a spectral representation for the bosonic field renormalization constant, which in QED (where the boson is the photon), doubles as the charge renormalization constant:

$$Z_3^{-1} = 1 + \int_0^\infty \rho(\kappa^2) d(\kappa^2) \qquad (24)$$

This was the first nonperturbative expression for a renormalization constant. It involves a slight redefinition of the spectral function ρ of the electromagnetic field. Since the spectrum of the electromagnetic potential will always include a massless ($\kappa^2 = 0$) one-photon state, the corresponding spectral function will always include a summand $\delta(\kappa^2)$. In equation 24, this summand has been separated off, resulting in the first term on the right-hand side. The spectral function in the second term then only involves states with two or more particles. We will be using this definition of the spectral function from now on.[105]

The point of this redefinition is to establish explicit numerical bounds on the renormalization constant. This also requires invoking the positive definiteness of the spectral function. Due to the presence of negative-norm states with scalar photons in the Gupta–Bleuler formalism, the positive definiteness of ρ in QED was not quite as immediately evident as for a scalar. However, as Källén showed, positive definiteness ultimately resulted from the usual cancellation between contributions from scalar and longitudinal photons.[106] From equation 24 and the positive definiteness of ρ, Källén could then conclude that Z_3 must lie between 0 and 1.

---

[105] An analogous redefinition of the spectral function can of course be done for massive fields as well, by separating off the summand $\delta(\kappa^2 - m^2)$ corresponding to the massive one-particle state. For the massive case, the only difference in equation 24 is then that the lower bound of the integral in the second term on the right-hand side is determined by the lowest energy state *above* the one-particle state. This lower bound will be greater than m^2 for $m^2 > 0$ (mass gap).

[106] It is this argument that fails for non-abelian gauge theories, where the bare charge can be smaller than the renormalized one (asymptotic freedom).

Källén had thereby proved nonperturbatively that, no matter what kind of charged matter one assumed, charge renormalization would always only lead to a reduction of the bare charge (Källén, 1952, p. 426). This result had first been derived by Schwinger, although he does not appear to have published it. It was also derived independently of Källén in Japan, in the group of Shoichi Sakata (Umezawa and Kamefuchi, 1951). And it was a very important result, which Pauli, in particular, had always placed great emphasis on.[107] It showed that the effects of charge renormalization could never be reproduced by heavy auxiliary fields. For mass renormalization, one could introduce heavy scalar fields and couple them to the electron. They gave an infinite contribution to the self-energy that had the opposite sign as compared to the electromagnetic self-energy, leading to a compensation of the two infinite effects, at least to some order in perturbation theory. This mechanism had formed the basis of several models that anticipated results of renormalization theory through the cancellation effects of heavy fields (Blum, 2015, pp. 87–88), in particular by Sakata (1947). However, Källén's theorem proved that no such compensation was possible for charge renormalization. This made a realistic interpretation of charge renormalization (where the necessary subtractions are understood as the effects of an actual physical field) impossible. Kamefuchi recalls that this result managed "to kill his [Sakata's] method and philosophy."[108]

4.2 Källén's Proof of Infinite Renormalization

Källén's result left open a central question: was Z_3 actually a finite number or was it infinitely small, having to compensate an infinitely large bare charge as implied by perturbation theory? In other words – and this finally brings us to the question with which we started out this chapter – was the renormalization constant Z_3^{-1} really infinite, independent of perturbation theory? Källén (1953) engaged with this question in detail in a paper penned after his return to Sweden from Zurich. As we shall see, he was ultimately unable to make any definitive statements about charge renormalization specifically, only about the entire set of renormalization constants in QED: the inverse charge renormalization constant Z_3^{-1}, the mass renormalization constant δm, and the inverse square root of the electron wave function renormalization constant $Z_2^{-1/2}$. All three of these constants were infinite in perturbation theory, and Källén sought to determine whether they would remain so in his nonperturbative formulation of QED.

---

[107] Letter from Pauli to Matthews, December 14, 1949 (von Meyenn, 1993, p. 720).
[108] Fax from Kamefuchi to the author, June 29, 2022. Kamefuchi went on to remark that Sakata "was content with the fact that his method played a pioneering role in the historical development of the [renormalization] theory."

To this end, he devised a relatively involved *reductio* proof, showing that, under the assumption that all three renormalization constants were finite, one could obtain a constant lower bound for the spectral function ρ of the photon. This implied that the integral in equation 24 would not converge, rendering Z_3^{-1} infinitely large. This was in contradiction with the original assumption, thereby implying that at least one of the renormalization constants must be infinite.

The lower bound for the spectral function was obtained by considering only a select subset of states in equation 23. As each of these states gives a positive contribution to the spectral function, this indeed leads to a lower bound. The specific states that Källén considered were states $|q,q'\rangle$, which included only an electron–positron pair with momenta q and q'. These are the simplest many-particle states possessing the same quantum numbers as the photon. This simplicity allowed Källén to obtain a nonperturbative expression (in terms of unknown spectral functions) for the relevant matrix elements. The resulting expression was then analyzed in the limit of large total momentum $q + q'$. In this limit, several terms dropped out under the assumption that all renormalization constants were finite (and thus that the relevant spectral functions went to zero in the limit). In this manner, Källén obtained the following relation for the relevant matrix elements of the current operator:[109]

$$\lim_{(q+q')^2 \to \infty} \langle 0|j_\mu|q,q'\rangle = \langle 0|j_\mu^{\text{Born}}|q,q'\rangle \frac{2\sqrt{Z_2} - 1}{Z_3} \tag{25}$$

where j_μ^{Born} is the renormalized Heisenberg current, but calculated only to leading order in perturbation theory (Born approximation). The right-hand side could be calculated explicitly and the left-hand side could then be inserted into the expression for the spectral function ρ, to obtain a lower limit for that function at large energies:

$$\lim_{\kappa^2 \to \infty} \rho(\kappa^2) > \frac{e^2}{12\pi^2 \kappa^2} \left(\frac{2\sqrt{Z_2} - 1}{Z_3} \right)^2 \tag{26}$$

If this lower limit is greater than zero, then equation 24 implies that Z_3^{-1} is logarithmically infinite. Consequently, the right-hand side must be zero. The denominator cannot become infinite, because Z_3 is less than 1. Therefore, the only remaining possibility is for the numerator to be zero. This seemed possible in principle, for $\sqrt{Z_2} = 1/2$. This exceptional case was somewhat irritating and was dismissed by Källén as too special to warrant serious consideration.

---

[109] Note that Källén gave the spectral function for the electromagnetic field in terms of matrix elements of the electromagnetic current operator by applying the Heisenberg equations of motion, apparently to avoid issues with gauge noninvariance.

As we shall see in chapter 4 of the second volume, this strange exception indeed disappeared when Källén slightly corrected his result a couple of years later.

Disregarding this transient anomaly, Källén had – starting from the assumption that all renormalization constants are finite – reached the conclusion that Z_3^{-1} is infinite. While this was *not* a proof that Z_3^{-1} was infinite, it *was* a *reductio ad absurdum* from which one could conclude that the initial assumptions were invalid, i.e., "that at least one (and probably all) of the renormalization constants is infinite." (Källén, 1953, p. 15)

Källén's proof had some shortcomings. For one, despite being a model of rigor when compared to, e.g., the proofs for the divergence of the perturbation series, it was far from unassailable, as Källén himself was the first to concede in his paper:

> The proof presented here makes no pretense at being satisfactory from a rigorous mathematical point of view. It contains, for example, a large number of interchanges of orders of integrations, limiting processes and so on. From a strictly logical point of view we cannot exclude the possibility that a more singular solution exists where such formal operations are not allowed. (Källén, 1953, p. 15)

However, the proof's most important drawback was its indirectness: it gave no indication of how the renormalization constant(s) would diverge in any actual calculation, nor which constants would diverge, nor what the physical interpretation of that divergence should be. For Källén, the proof appears to have been little more than a lemma of formal relevance. But others soon looked at the matter more directly and physically.

4.3 Gell-Mann, Low and the Ultraviolet Behavior of QED

There was a certain tension in how the renormalization constants were interpreted in the early 1950s. On the one hand, the actual value of these renormalization constants was considered a "purely academic question" (Thirring, 1951, p. 463). After all, they did not show up anywhere in the physical predictions of QED, just as little as the cutoff used to regularize the theory. On the other hand, renormalization was closely tied to a failure of QED at high energies or short distances – in the ultraviolet (UV), for short. Schwinger (1948) had characterized renormalization as a procedure that could "isolate those aspects of the current theory that essentially involve high energies, and are subject to modification by a more satisfactory theory, from aspects that involve only moderate energies and are thus relatively trustworthy." In principle, however, the renormalized theory could give predictions for electrodynamic processes at very high energies – all of the divergences had been removed, after all.

Of course at short distances the nuclear interactions would become important, even dominant. Consequently, no one had really investigated the UV behavior of QED – it was an empirically irrelevant question. But it was finally asked in the summer of 1953, by the already mentioned pair of Murray Gell-Mann and Francis Low, now no longer at the Institute for Advanced Study but together at the University of Illinois.[110]

According to Källén,[111] their work had originated from an attempt to give an inconsistency proof for QED of their own. In studying mass renormalization, they had discovered a quadratically divergent, non–gauge-invariant mass term for the photon, which they apparently briefly believed spelled trouble, until Källén convinced them that this was a problem well-known from perturbative calculations and had been solved through the gauge-invariant regularization technique of Pauli and Villars (1949). The paper that Gell-Mann and Low (1954) ultimately published, which is now primarily famous for the introduction of renormalization group methods, instead dealt with, as indicated in its title, "Quantum Electrodynamics at Short Distances."[112]

In the paper, they investigated whether QED could in principle be applicable at arbitrarily short distances. As the paradigmatic process for studying the short-distance behavior, they took the scattering of two electrons (Møller scattering), exchanging a four-momentum k^2 through a virtual photon. This four-momentum will be spacelike, i.e., $k^2 < 0$.[113] Investigating short distances thus meant probing the behavior of the photon propagator as $k^2 \to -\infty$. It should be stressed that this had not been done before. And, as Gell-Mann and Low highlighted, it was a question that could not really be addressed within perturbation theory. The reason for this are the so-called large logarithms. The n-th order term in the perturbative expansion of the renormalized propagators is dominated, for large $|k^2|$, by the term proportional to $\left(\ln k^2/m^2\right)^n$, with m the electron mass. This implies that higher-order contributions become increasingly important for large $|k^2|$. While this was not a problem for experimentally

---

[110] See interview with Gell-Mann by Sara Lippincott, p. 29, as referenced in footnote 103.

[111] Letter to Pauli of June 30, 1955 (von Meyenn, 2001, p. 291).

[112] On the significance of this paper in the early history of the renormalization group, see Fraser (2021). Li (2015, p. 1216) has emphasized the importance of the renormalization group "as a way not just to calculate empirical quantities but also to determine whether a given Lagrangian exists in the UV limit." Indeed, exploring the UV limits of QFTs was precisely what Gell-Mann and Low used their newly developed renormalization group methods for in this pioneering paper.

[113] This is nowadays called t-channel scattering, where the momentum squared of the virtual photon k^2 is given by the square of the four-momentum transfer, the Mandelstam variable t, and is thus negative. For processes like electron-positron scattering, there is also s-channel scattering, where k^2 is given by the square of the center-of-mass energy, the Mandelstam variable s, and is thus positive (timelike virtual momentum).

accessible scattering processes, at some point these logarithms would start cancelling out the increasing powers of the coupling constant. Specifically, perturbation theory would break down completely when

$$\ln k^2/m^2 \approx 1/e^2 \tag{27}$$

This implied distances far smaller than the size of the nucleus. Before Gell-Mann and Low's work, no one had considered it worthwhile to ask how QED would behave at such short distances. Even Dyson, at his most optimistic, had (often explicitly) assumed that *any* quantum field theory would be insufficient at very small distances; he had merely believed that this insufficiency would not yet be manifest at nuclear scales. Despite the success of renormalization theory, it was still generally accepted that QFT would not be able to describe the ultimate structure of matter. This entrenched view, which was to some extent a holdover from the long pre-renormalization crisis of QFT, was first challenged by Gell-Mann and Low. They thus expanded the scope of possible outcomes for probing the consistency of QED. Until then, the only options had been full inconsistency (as Källén was angling for) or a more or less consistent theory that in some manner (which no one had really investigated) would become unreliable, and possibly even divergent, at very short distances. To these two options, Gell-Mann and Low added a third: QED might, at least in principle, be a final theory. They were, of course, fully aware that QED by itself was empirically inadequate as a final theory candidate. But it could serve as a blueprint for one:

> [T]he first few terms [in a perturbation expansion of QED] should suffice for calculation unless r is as small as $e^{-137}\hbar/mc$, a ridiculously small distance. We have no reason, in fact, to believe that at such distances quantum electrodynamics has any validity whatever, particularly since interactions of the electromagnetic field with particles other than the electron are ignored. However, a study of the mathematical character of the theory at small distances may prove useful in constructing future theories. Moreover, in other field theories now being considered, such as the relativistic pseudoscalar meson theory, conclusions similar to ours may be reached... (Gell-Mann and Low, 1954, p. 1301)

How did they go about this? The process whose UV behavior Gell-Mann and Low studied was the scattering of two charged particles. In the Green's function approach, this process is determined by the photon propagator D_{FR}. At first order in perturbation theory this is equal to the free propagator $1/k^2$, which is the Fourier transform of the Coulomb potential $1/r$. As Gell-Mann and Low pointed out, the most straightforward way to physically interpret the modified propagator was then to read it as radiative corrections to the electrostatic interaction. While these corrections to the electrostatic potential due to vacuum polarization had been known for a long time (Uehling, 1935), Gell-Mann and Low were the first to describe the short-distance features of the potential through scattering processes with large momentum transfer and thus through the values of the photon propagator D_{FR} for large off-shell momenta $|k^2|$.

In order to study the UV behavior of D_{FR}, Gell-Mann and Low used spectral decomposition methods similar to those developed by Källén and Lehmann. While Källén had only derived a spectral representation for the VEV of the commutator of two fields, both Lehmann and Gell-Mann and Low had derived an analogous spectral representation for the two-point Green's functions, the propagators. For a massless bosonic field, the spectral representation of the renormalized propagator D_{FR} is given by

$$D_{FR}(x,x') = D_F(x,x') + \int_0^\infty D_F(x,x';\kappa^2)\rho(\kappa^2)d(\kappa^2) \tag{28}$$

where $D_F(x,x';\kappa^2)$ is the free propagator of a field with mass κ^2, and the spectral function ρ is the same as in Källén's representation. In momentum space, this becomes

$$D_{FR}(k^2) = \frac{1}{k^2} + \int_0^\infty \frac{1}{k^2 - \kappa^2 + i\epsilon}\rho(\kappa^2)d(\kappa^2). \tag{29}$$

This was the quantity whose behavior Gell-Mann and Low wanted to study for large $|k^2|$. It is here that they established the connection between their study of UV behavior and Källén's work on the renormalization constants. Taking the limit of this propagator for $k^2 \to -\infty$, Gell-Mann and Low obtained

$$\begin{aligned}\lim_{k^2 \to -\infty} D_{FR}(k^2) &= \lim_{k^2 \to -\infty}\left(\frac{1}{k^2} + \int_0^\infty \frac{1}{k^2 - \kappa^2}\rho(\kappa^2)d(\kappa^2)\right)\\ &= \frac{1}{k^2} + \int_0^\infty \lim_{k^2 \to \infty}\frac{1}{k^2 - \kappa^2}\rho(\kappa^2)d(\kappa^2)\\ &= \frac{1}{k^2} + \frac{1}{k^2}\int_0^\infty \rho(\kappa^2)d(\kappa^2)\\ &= \frac{1}{k^2 Z_3}.\end{aligned} \tag{30}$$

using the spectral representation of the renormalization constant, equation 24. The essential element here is, of course, taking the limit in the integral; strictly speaking, this can only be done if the spectral function ρ falls off quickly enough (Matsubara et al., 1984, p. 873f). In the limit of infinite off-shell momenta, the renormalized photon propagator is then the propagator of a free, massless field multiplied by Z_3, the charge renormalization constant. And Gell-Mann and Low provided a concrete physical interpretation of this mathematical result: the electrostatic potential close to the origin is again a pure Coulomb potential, but now for a charge of magnitude $Z_3^{-1}e$. And this was indeed to be expected: close to the particle, one saw the unmodified Coulomb potential of the bare charge. This was the most physical interpretation of the bare charge since its introduction several years earlier. It also meant that, in studying the UV behavior of QED, Gell-Mann and Low were effectively investigating the magnitude of the charge renormalization constant.

To study this UV behavior, Gell-Mann and Low introduced a novel regularization scheme, where the unrenormalized modified propagators are regularized such that they are equal to the free propagator (with the physical mass) at a momentum k with $k^2 = -\lambda^2$ (Wilson, 1971, pp. 1843ff). Formally, this new regularization method is equivalent to a modern renormalization condition with a renormalization scale λ. However, one needs to be careful here, because Gell-Mann and Low did not interpret the propagator fixed by this condition as already renormalized. To them, imposing the "renormalization" condition at the scale λ was merely an intermediate step, which they considered to be a form of *regularization*; *renormalization* was to be performed in an additional next step by enforcing the usual on-shell renormalization condition. Consequently, Gell-Mann and Low considered λ to be a regularization scale, that is, a form of cutoff.[114] What we would now consider a propagator renormalized off the mass shell, Gell-Mann and Low viewed merely as a regularized, and still unrenormalized propagator. The cutoff of Gell-Mann and Low was interpreted as a renormalization scale only later by Bogoliubov and Shirkov (see part 2, chapter 3).

In contrast to their innovative regularization method, Gell-Mann and Low were more traditional in their reliance on perturbation theory. They worked under the assumption that if a certain UV behavior was shared by all terms (to arbitrarily high order) in a perturbative expansion of the propagator, then the full propagator itself would exhibit the same UV behavior – even if, as was to be assumed, the full propagator was not the sum of the perturbation series. This assumption was particularly useful for the regularized propagator, which contained three distinct scales, k, λ, and m, the electron mass, making the UV limit nontrivial. Gell-Mann and Low proposed to neglect m in the limit where k and λ were large. In general, one could not simply set $m = 0$, as doing so might lead to infrared divergences. However, when the regularized propagator is calculated in perturbation theory, such divergences do not arise. Gell-Mann and

---

[114] This interpretation of Gell-Mann and Low's "cutoff" scale is discussed in Fraser (2021, p. 121). In addition to their study of the photon propagator, Gell-Mann and Low also investigated the UV behavior of the electron propagator in an approximation where the photon self-energy is neglected. In this part of their paper (section 3), they used a more traditional regularization procedure, where the regularization scale λ can straightforwardly be interpreted as a UV cutoff.

Low now assumed that this feature carried over to the nonperturbative theory, where the UV limit of the full regularized propagator could then be obtained by setting the electron mass equal to zero. Their approach was criticized by Källén in a short review for the *Zentralblatt für Mathematik* (Källén, 1955), where he argued that a series does not necessarily inherit the functional properties of its individual summands, even if that series converges.

The role of perturbation theory in Gell-Mann and Low's approach would be reevaluated in the 1960s by Kenneth Wilson, and we will revisit this point in volume 2, chapter 5. However, already in 1954, it was clear that Gell-Mann and Low were able to obtain results that appeared to transcend perturbation theory. By extrapolating the properties of the terms in the perturbation series, they could derive a functional equation for the renormalized photon propagator, parameterized as

$$D_{FR} = \frac{1}{k^2} d_R(k^2/m^2). \tag{31}$$

They found that the most general solution to their functional equation was given in terms of two unknown functions ψ and q. These two functions, which remained undetermined by the functional equation, were related to the solution d_R in a somewhat unusual way, through the following equation:

$$\ln(-k^2/m^2) = \int_{q(e^2)}^{e^2 d_R(k^2/m^2)} \frac{dx}{\psi(x)} \quad (-k^2 \gg m^2). \tag{32}$$

This equation is a little hard to read. The essential point is that the left-hand side diverges as k^2 goes to $-\infty$, so the right-hand side must, too. The question then is whether d_R also has to go to infinity for that to happen. The UV limit of d_R is the charge renormalization constant, so this would imply that the bare charge is infinite and QED would be divergent at short distances. For the integral to diverge without d_R having to go to infinity, the integrand must diverge at a finite value of x, that is, the function ψ must have a zero for some finite x_0. In this case, $e^2 d_R$ would only need to attain the value x_0 as k^2 goes to $-\infty$ in order to ensure that the left-hand side diverges. x_0, the location of the zero of ψ, could then be identified with the (finite) value of the bare charge. Quite strikingly, this value turned out to be entirely independent of the observed physical charge.

The solution to the Gell-Mann–Low functional equation thus left open the possibility that the bare charge might be nonperturbatively finite. This was not, as Gell-Mann and Low pointed out, in conflict with Källén's result that one of the renormalization constants was infinite; it would then just *not* be the charge renormalization constant. But Gell-Mann and Low had provided a new and direct nonperturbative handle on the magnitude of the renormalization constants, which would prove to be very important. For the time being, however, it provided only a minor and possibly entirely inconsequential qualification of Källén's result. In any case, it was still not at all evident that any sort of statement about the magnitude of the renormalization constants was relevant to the question of the consistency of QED. The magnitude of the charge renormalization constant, at least, had become somewhat less academic, since it appeared even in the renormalized theory, in the UV limit of the photon propagator.

But what would it mean if QED had solutions that diverged at short distances? Gell-Mann and Low appeared to believe that a final theory had to be finite in the UV. This question was not debated further at the time.[115] Which is curious, because it is not at all clear how the argument that an infinite bare charge implies the inconsistency of the theory would actually unfold.

After all, the charge itself is not directly observable; it needs to be extracted from a scattering cross section, for example, for Gell-Mann and Low's prime example, Møller scattering. And the high-energy behavior of that scattering cross section will not solely depend on the UV behavior of the photon propagator (and thus on the bare charge), but also on the UV behavior of the vertex function (which modifies the tree-level graph) or of the four-point function (which gives the contribution from the irreducible box diagram). One could choose to ignore these complications, and simply take the cross section for Møller scattering calculated at leading order in perturbation theory (Jauch and Rohrlich, 1955, section 12–1):

$$\frac{d\sigma}{d\Omega} = \frac{e^4}{m^2 E^2} X \tag{33}$$

where E is the energy of one electron in the center-of-mass system, and X is the spin sum, which depends only on the scattering angle in the $E \to \infty$ limit. But if one takes this limit, one sees that the behavior of the cross section essentially depends on how the charge goes to infinity. If it goes as $\ln E^2/m^2$, the cross section in this approximation will in fact remain finite for all energies.[116]

---

[115] Nor, it should be mentioned, is it debated today. It would seem that the possibility of a theory being asymptotically infinite is considered largely irrelevant. Physicists are happy to consider only the two extreme cases of UV behavior: the Landau pole (clearly inconsistent) and asymptotic freedom (plausibly consistent). This is also due to the fact that while these extreme cases are instantiated by prominent QFTs (QED and QCD, respectively), there is no paradigmatic example of an asymptotically infinite theory. Or rather: no longer. In 1954, asymptotic infinity was, of course, still considered the most plausible UV behavior for QED.

[116] This means, a fortiori, that it will not come into conflict with the unitarity bounds on the total cross section, which were discovered several years later (Froissart, 1961).

Neither Gell-Mann and Low nor Källén thus provided immediate insight into the question of the consistency of QED. Rather, the upshot of their work was that the infinities of QED would not simply vanish in a nonperturbative analysis. New solutions that did not rely on a perturbation expansion and did not involve divergent quantities – solutions that would unambiguously establish the consistency of the theory – were not on the horizon. Those who still felt that the consistency of QED was a worthwhile question had to look for alternative ways of investigating it. We will explore these other approaches in the following sections, starting with arguably the most radical one: the axiomatic approach.

5 The Axiomatic Approach

The story of axiomatic QFT is quite distinct from the story reconstructed in the other parts of this Element. First, it is a lot closer to mathematics proper. In stark contrast to the intricate manipulations and approximations of middlebrow nonperturbative QFT that we have seen so far, we will in this section have to engage with highbrow papers written in the theorem-proof style of mathematics and mathematical physics. Second, the story of axiomatic QFT cannot be read as a prelude to the discovery of asymptotic freedom. Axiomatic QFT would go off on its own trajectory, ultimately becoming an established subdiscipline of mathematical physics. We will not be able to follow it very far along this trajectory. Instead, we will merely follow it from the origins of axiomatic QFT in the 1930s to its first major contribution to the consistency debate: the famous theorem formulated by Rudolf Haag in the mid 1950s.

The origins of axiomatic QFT can be traced back to a tradition associated with the University of Göttingen, and more specifically with David Hilbert, of studying, "by means of axioms, those physical sciences in which mathematics plays an important part" (Browder, 1976, p. 14). Hilbert first formulated this program in 1900 as the sixth of a set of 23 problems he deemed central to the development of mathematics in the twentieth century. His subsequent work on the axiomatization of physics led him from the established theories of classical physics to general relativity and, ultimately, to quantum mechanics (Corry, 2004). Hilbert's work on QM was conducted in collaboration with Lothar Nordheim and John von Neumann and was the immediate predecessor to von Neumann's foundational textbook *Mathematische Grundlagen der Quantenmechanik* (v. Neumann, 1932). Von Neumann's book, in turn, became the starting point for axiomatic QFT. Attempting to give an axiomatic formulation of QFT – with all its foundational issues, which the reader will by now be familiar with – would have been well-nigh impossible had physicists not been able to piggyback on von Neumann's axiomatic formulation of quantum mechanics.

What did the axiomatic approach change with regard to the question of consistency? It certainly implied an increase in mathematical precision, though the axioms were never cast in a formal language and remained "soft" in the sense of Rédei and Stöltzner (2006). But the new approach did not simply result in a more rigorous formulation of the axioms we have already discussed, namely, (A) the basic tenets of quantum mechanics and (B) the dynamical equations of QED. As I have argued, these were the implicit axioms underlying the consistency debate up to this point. One might have imagined an axiomatic approach to QFT based on making these axioms explicit and precise. After all, Hilbert had readily included dynamical equations in his axiomatic systems, either directly – Newton's equations with specific force laws in classical mechanics[117] – or as immediate corollaries of the axioms – Hilbert action and Einstein equations in general relativity (Hilbert, 1915).

But in the axiomatic approach to quantum mechanics, the dynamical equations were of reduced importance. For von Neumann, the state space was primary, and his axiomatics was an axiomatics of Hilbert space. The time-dependent Schrödinger equation and the special dynamical role of the Hamiltonian were introduced as part of the historical introduction (von Neumann, 1932, p. 8), not as part of the mathematical foundation. Early axiomatic QFT followed von Neumann's approach in this regard. Consequently, it did not deal with the physical details of quantum electrodynamics but rather with the formal aspects of an abstract quantum field theory. But without the field equations, what was the decisive factor that set this abstract QFT apart from standard quantum mechanics?

Von Neumann's book had nothing to say on this point. Its treatment of QFT was rudimentary, amounting to little more than a paraphrase of Dirac's foundational paper on radiation theory (Dirac, 1927b). But over the course of the 1930s, von Neumann and his compatriot, friend, and collaborator Eugene Wigner explored the possibility of a similarly Hilbert-space-based, axiomatic approach to QFT. Von Neumann's approach was to loosen the axioms of quantum mechanics to encompass systems with an infinite number of degrees of freedom, while Wigner supplemented (A) the basic tenets of quantum theory with an additional axiom, namely (B') Poincaré invariance. Von Neumann thus approached QFT as quantum *field* theory, whereas Wigner viewed QFT as relativistic quantum theory.

---

[117] In his 1905 lectures on mechanics; see (Corry 2004, section 3.3.1).

The work of both Wigner and von Neumann was highly influential in shaping the post-renormalization axiomatic QFT of the 1950s, although it was Wigner's vision of relativistic quantum theory that ultimately came to define axiomatic QFT most strongly.[118] This emphasis on Poincaré invariance, rather than on the specific dynamical laws of QED, made the debate on consistency within this tradition quite distinct from the debates discussed elsewhere in this Element – it was quite literally the consistency of a different system of axioms that was at stake. This difference was ultimately far more relevant than the fact that one tradition made its axioms more explicit than the other.

In the next two sections, we will discuss the work of von Neumann and Wigner in the 1930s. These early explorations were highly tentative and did not directly engage with the divergence difficulties that dominated the discourse on QFT during that period; as Wigner later admitted, he "was a little afraid that the problems of field theory were not ripe for solution."[119] Yet, taken together, von Neumann's work on systems with an infinite number of degrees of freedom and Wigner's work on Poincaré-covariant Hilbert spaces formed the basis for the formulation of Haag's theorem, which we will discuss in Sections 5.3 and 5.4. Wigner provided the immediate impetus and the framework for the formulation of axiomatic QFT, while von Neumann was the first to clearly identify mathematical features that distinguished QFT from quantum mechanics and that ultimately stood in the way of a straightforward realization of the axiomatic program.

5.1 Von Neumann and Nonseparable Hilbert Spaces

The first in-depth study of the state space of QFT was not performed by von Neumann but by Vladimir Fock (1932) in the Soviet Union. Fock identified this state space as a "sequence of configuration spaces for 1, 2, ... etc. particles." While the space thus constructed now carries Fock's name, he did not actually analyze its mathematical structure in much detail. He clearly appears to have assumed that there was no qualitative difference between the state space of quantum mechanics and that of QFT. In particular, he tacitly assumed that his space was not supposed to contain states with an infinite number of particles, that is, the limit of his sequence of configuration spaces.

---

[118] The contrast between an axiomatic QFT that views Poincaré invariance as central and a field-theoretic approach to QFT that treats Poincaré invariance as a negotiable feature of the dynamics can still be observed in the twenty-first century (Fraser, 2016, chapter 5).

[119] Interview with Thomas S. Kuhn, December 14, 1963, www.aip.org/history-programs/niels-bohr-library/oral-histories/4963-3.

Around the same time, von Neumann was collaborating with Pascual Jordan, a pioneer of QFT. Jordan had voiced his opinion that a functioning theory of quantum fields – one that avoided the divergence difficulties – would require a generalization of the mathematical framework of quantum mechanics (Jordan, 1933). Together with Jordan (and Wigner), von Neumann explored the possibility of generalizing the structure of the quantum mechanical operator algebra.[120] This collaboration likely inspired von Neumann's investigations into other generalizations of QM, specifically of its Hilbert space structure. He thus began exploring the consequences of modifying the axioms he himself had introduced for a Hilbert Space $\mathcal{H}$.

One of these axioms was that of "separability", which essentially stipulated that any orthonormal set in $\mathcal{H}$ (and thus any orthonormal basis) is *at most* countably infinite. Another axiom ensured that any complete orthonormal set – that is, any orthonormal basis – is *at least* countably infinite.[121] In a lecture series given in Princeton in the academic years 1933/34 and 34/35,[122] von Neumann identified these two axioms as the only two that could be sensibly modified to construct a more general theory of "linear spaces" that included the theory of Hilbert spaces. Relaxing (or rather: negating) the axiom of infinitude simply gave the well-known finite-dimensional Euclidean vector spaces $\mathbb{C}^n$. But negating the axiom of separability produced a new mathematical entity, which von Neumann later sometimes referred to as a hyper-Hilbert space (Jordan and von Neumann, 1935). Following the later custom, I will be referring to this entity as a nonseparable Hilbert space.[123]

The pièce de résistance of the lecture series was von Neumann's proof that any linear space had an orthonormal basis, without having to make use of the axiom of separability – though making essential use of the axiom of choice (Theorem 12.19). For nonseparable Hilbert spaces, this orthonormal basis would be uncountable; however, even in such spaces, any element could

---

[120] On Jordan's program of an algebraic generalization of quantum mechanics, which spanned several decades, see Liebmann et al. (2019), arXiv:1909.04027.

[121] In (von Neumann, 1929, p. 65) these are labelled as axioms C and D. Confusingly, the labeling is reversed in (von Neumann, 1951, p. 20). And in a footnote in Jordan and von Neumann (1935), separability is (in what is presumably a typo) referred to as axiom E.

[122] Reprinted as von Neumann (1951).

[123] Von Neumann also showed that any nonseparable Hilbert space would obey the same axioms, regardless of the exact uncountable cardinality $\aleph$ of its basis (p. 31). Later this could be taken to imply that the only qualitatively relevant transition is that from quantum systems with a finite number of degrees of freedom to quantum systems with an infinite number of degrees of freedom. This transition corresponds to the transition from a countably infinite to an uncountably infinite basis for the Hilbert space. By contrast, the transition from quantum systems with a countably infinite number of degrees of freedom to those with an uncountably infinite number of degrees of freedom would not introduce any fundamentally new structures.

still be written as a linear combination of only a countable number of basis elements (p. 29). Von Neumann thus concluded that "these spaces are very similar to Hilbert Space with regard to most of their important properties. This is due to the fact that every separable part of them lies in a subset of them which is a [closed, separable] Hilbert Space" (p. 32–33).

The appendix to von Neumann's lectures contains an example of a nonseparable Hilbert space (example 3, pp. 46–48) that bears a striking resemblance to fermionic occupation number space: it is the space of complex-valued, finite-norm functions (the wave functions) defined on the direct product of a countably infinite number of spaces (corresponding to the one-fermion states), each containing exactly two linearly independent elements. These elements correspond to the occupation numbers zero and one, and von Neumann even calls these elements 0 and 1. This should not be overestimated, since the use of 0 and 1 is also typical in the construction of a power set. And this is what von Neumann was constructing here, the Hilbert-space analog of a power set – arguably the simplest possible nonseparable Hilbert space. But by the time he gave his first and only seminar series explicitly on QFT, one year later, in the academic year 1935/36,[124] he most certainly had established the connection between his model of a nonseparable Hilbert space and fermionic occupation number space.

For us, this raises the question: who was right? Fock, who considered his space a regular Hilbert space, or von Neumann, who identified the state space of QFT as nonseparable? The answer is (of course?) that they were both right. Fock's space was a regular Hilbert space, but it differed from von Neumann's in that it excluded states with an infinite number of particles. Fock had ignored these states when demonstrating the equivalence of his space with occupation-number space. Von Neumann (1939), in his last paper on the subject, clarified the distinction between the two spaces: he considered infinite direct products $\prod \otimes_\alpha \mathcal{H}_\alpha$, where each $\mathcal{H}_\alpha$ is a regular Hilbert space. He could show that such a nonseparable Hilbert space (a "complete direct product") could be decomposed into mutually orthogonal subspaces (the "incomplete direct products," p. 35). If a general element of the complete direct product was written as $\prod \otimes_\alpha f_\alpha$ (with f_α an element of $\mathcal{H}_\alpha$), then the elements of a given incomplete direct product differed from each other only in a finite number of f_α (p. 24, lemma 3.3.5). The identification of Fock's space – whose elements only differed by a finite number of particles – with an incomplete direct product was tempting, but von Neumann merely remarked:

---

[124] Reprinted in Rédei and Stöltzner (2001, pp. 249–268).

> What happens could be described in the quantum-mechanical terminology as a "splitting up" of $\prod \otimes_\alpha \mathcal{H}_\alpha$ into "non-intercombining systems of states", corresponding to the "incomplete" direct products [...]. This viewpoint, as well as its connection with the theory of "hyperquantization" [i.e., second quantization], will be discussed elsewhere. (p. 4)

But this never happened; unlike Wigner, von Neumann did not return to quantum field theory after the war, instead devoting himself to the development of computing machines. The question thus remained unanswered: what was the state space of QFT? It undoubtedly contained an infinity of one-particle states. Normally these were discretized by putting the system "in a box," which meant that there was a distinct lowest-energy state, eliminating the problem of infrared divergences. But this still left a countable infinity of one-particle states extending to infinitely high energies. Should the state space include states with an infinite total number of particles occupying these one-particle states? Such states with infinite total particle number did not appear as intermediate states in perturbation theory calculations. They could also be removed through a cutoff procedure (as von Neumann effectively did in his QFT seminars – and with him many others), because there was then only a finite number of one-particle states to occupy. So, all in all, the states with infinite particle number seemed largely irrelevant, and von Neumann's assessment that nonseparable linear spaces were "very similar to Hilbert Space with regard to most of their important properties" seemed vindicated by the practice of QFT. While von Neumann's forays into quantum field theory may thus well have seemed sterile at the time, his results would become relevant more than a decade later. However, it was Wigner's attempt at analyzing the Hilbert space structure of QFT from the perspective of relativity that would form the actual starting point of axiomatic QFT.

5.2 Wigner, Wightman and Relativistic Quantum Theory

In the late 1930s, Wigner (1939) presented a paper on relativistic quantum theory that deliberately avoided engaging with the concerns of the quantum field theory of the day, the properties of fields, the dynamical interactions, the divergences, or even the infinite number of degrees of freedom. Instead, it focussed exclusively on the transformation properties of the states in the Hilbert space of a relativistically covariant quantum theory. By taking the Hilbert space as a starting point, rather than a specific dynamics or set of variables, Wigner's approach clearly echoed von Neumann's original approach to quantum mechanics.

Wigner managed to construct all unitary irreducible representations of the Poincaré group. He further showed that in fact every unitary representation of the Poincaré group could be decomposed into irreducible representations, a nontrivial statement for an infinite group. With these unitary representations in hand, he could then construct the Hilbert space. If all transition probabilities (i.e., the scalar product in the Hilbert space) were to remain invariant under Poincaré transformations – as was to be expected in a relativistic quantum theory – then all states in the Hilbert space must transform under such a unitary representation. As a result, he could construct a Hilbert space for a relativistic quantum theory simply by enumerating the different unitary representations of the Poincaré group that comprised it.

At the time, Wigner did not delve into the physical interpretation of the Hilbert space thus constructed. But after the war, he revisited his results and developed them into a program for constructing relativistic quantum theory from the ground up – not by taking a classical field theory and then quantizing it, but by constructing a Hilbert space from the demands of relativistic invariance alone. This new perspective was first outlined in Wigner (1948).

Already at this stage, the advantages and disadvantages of the new viewpoint were apparent. On the positive side, focusing on the irreducible representations meant a huge decluttering compared to the field-theoretical viewpoint. For example, from the representation-theoretic perspective (Wigner called it the *invariantentheoretische Weg*), it was easily established that a representation characterized by a null energy-momentum four-vector (i.e., corresponding to a field theory with massless quanta) would only involve two spin degrees of freedom. This result required an elaborate derivation in field theory, where one started, e.g., for the case of spin one, with a four-component vector potential and had to eliminate all the unphysical and nondynamical degrees of freedom. The states in an irreducible representation were directly labeled by the properties of particles (energy-momentum and spin orientation/polarization), without reference to the possibly redundant field description. But this frugality came with a price: The baroque field-theoretical description provided a direct prescription for constructing invariant interaction Hamiltonians. By contrast, the representation-theoretic view, as Wigner clearly stated, gave "no indication on how to solve the problem of interaction."

Wigner explored the possibility of introducing interactions into his framework together with his PhD student, the Ukrainian-born Mexican physicist Marcos Moshinsky (1949). Moshinsky tried to extend Wigner's irreducible (one-particle) representations to construct covariant two-particle scattering states for two scalars or two Dirac fermions. These states were constructed as linear combinations of incoming plane waves and outgoing spherical waves. The question was now how to determine the relative amplitude between incoming and outgoing waves, that is, the scattering amplitude. In quantum mechanics, this was done by inserting the scattering ansatz into the Schrödinger equation with the interaction Hamiltonian. Instead, Wigner and Moshinsky attempted to determine the scattering amplitude by imposing further conditions on the scattering state. Specifically, Moshinsky imposed (boundary) conditions on the scattering wave functions at those points in many-particle configuration space where the coordinates of the two scattering particles coincide. These boundary conditions were motivated solely by demands of relativistic invariance, with some concessions to simplicity. Moshinsky's construction did not go very far. It allowed, for example, only for the scattering of spherical waves – that is, for collisions where the colliding particles have no relative angular momentum – and was never published. In general, Wigner himself was very cautious about promoting his approach to relativistic quantum theory. It was the young American Arthur Wightman who turned Wigner's particle perspective into a full-fledged research program.

Wightman initially wanted to do his PhD with Wigner "on the mathematical foundations of field theory," but, as Wightman recalled, Wigner warned him that this topic was "too hard."[125] Given that Moshinsky was working on related problems at the same time (Wightman eventually also graduated in 1949), one can only assume that Wigner objected to Wightman's frontal approach.[126] Wightman then ended up doing his PhD on a topic in meson physics with Robert Marshak and John Wheeler (MIW, p. 28). But shortly after graduating, he became the main proponent of Wigner's approach to QFT. In September 1951,[127] he came to the Niels Bohr Institute in Copenhagen for a sabbatical "and he carried with him from Princeton the wisdom of Wigner" (Haag, 2010b). Indeed, in Copenhagen he appears to have begun pursuing the Wignerian program for QFT in earnest, as evidenced by a letter to Fritz Rohrlich:[128]

> As far as my own work is concerned I am exceedingly optimistic. Wigner's methods are so powerful in field theory that I am getting blisters on my hands trying to use them. If I weren't so stupid slow and inaccurate, field theory would already have reached the stage in which divergences are of historical interest only. That is one of the reasons why I will be glad to be back in Princeton. I hope that with you, [Valentine] Bargmann and Wigner taking an interest the thing will go much further. I'll make you a bet that if you three do take an interest, the renormalization program will be dead as a doornail in June 1953. Wanna bet?

---

[125] Interview with Jagdish Mehra, January 18, 1988, JMP, Box 104, Folder 38, page 2. From now on, this interview will be referred to as MIW.
[126] Wightman recalls that he "forced [himself] on [Wigner]" (MIW, p. 32).
[127] See the Copenhagen Minute Book in the AHQP.
[128] The letter is undated, but sent from Copenhagen. It is in the Fritz Rohrlich Papers.

In particular, he later recalled that he pursued:

> ... the question that Wigner had left open, which I was very uncertain about mathematically, and which occupied an inordinate amount of time during my year in Copenhagen, was whether there is any physics in the unitary equivalence class of the reducible representation of the Poincaré group. I kept worrying that somewhere in the multiplicity theory there was some hint of an S matrix or something like that. (MIW, p. 5)

This was the problem that also Wigner and Moshinsky had pursued: whether the many-particle representations constructed from Wigner's original irreducible one-particle representations somehow contained the dynamics of the theory. Wightman presented his work at the previously mentioned inaugural CERN conference in June 1952. Programmatically, Wightman's presentation (Wightman, 1952) was more assertive than anything Wigner had ever said, explicitly referring to the "present difficulties for relativistic quantum theories" and characterizing his program not as a novel approach to QFT but as an attempt to build "up logically the requirements which must be imposed on *any* theory by the demand of invariance with respect to the inhomogeneous Lorentz group" (my emphasis). The results that Wightman presented, however, did not really go beyond what Moshinsky had done in his thesis three years earlier. It was becoming clear that the Wignerian program, based solely on a Poincaré-invariant Hilbert space, was not rich enough to produce the dynamics of quantum field theory. Further general principles were needed to sufficiently constrain the theory.

After Wightman's talk, Heisenberg remarked "that the theory of Wightman was in principle an S-Matrix theory" (p. 56). This remark, while not inaccurate, requires some unpacking. A decade earlier, Heisenberg had embarked on his own program of rebuilding QFT from scratch: the S-Matrix approach (Blum, 2017). This approach was indeed quite similar to that of Wigner and Wightman, as it too relied not on field dynamics but on implementing the constraints of relativity theory (Lorentz invariance of the S-Matrix) and of quantum theory (unitarity of the S-Matrix). Heisenberg had realized early on that these constraints were still too loose, an insight he shared with Wightman, remarking that "it would prove valuable to make another restriction of the possible theories in the scheme of Wightman" (p. 57). Heisenberg had explored several such further restrictions. The one he now suggested to Wightman had been introduced into the framework of S-Matrix theory by Ralph Kronig (1946): the demand of causality.

It was one thing to suggest adding causality as a further axiom to the Wigner–Wightman approach, but quite another to give a mathematical formulation of this demand. Kronig's idea had been to implement causality as a condition on the S-Matrix, where it appeared as Kramers–Kronig-type relations between the real and the imaginary parts of the S-Matrix elements. In this form, causality was indeed being explored in the early 1950s, not least by Wigner himself (Cushing, 1990, Section 2.6), although there is no indication that he connected this issue to his work on relativistic quantum theory.[129] By this time, however, Heisenberg no longer believed in a pure S-Matrix approach. When he recommended that Wightman incorporate the demand for causality into his program, he added that "it is difficult to enter on the question of causality without the concept of a field." This seems to be a reference to another version of the causality principle, first introduced by Pauli in his proof of the spin-statistics theorem (Blum, 2014) and nowadays known as "microcausality." Microcausality asserts that for a field operator ϕ,

$$[\phi(x), \phi(y)] = 0, \tag{34}$$

when the points x and y are spacelike separated. The two versions of causality – Kramers–Kronig analyticity conditions and microcausality – were related,[130] but they were formulated on very different levels. While Kronig's formulation pertained to scattering amplitudes, microcausality was concerned with the field operators.

There is no indication that Wightman took Heisenberg's remarks to heart. Upon returning to Princeton, he began putting together a book with Wigner and Bargmann, entitled *Foundations of Relativistic Quantum Mechanics*. A manuscript of the first few chapters of this unpublished book can be found in Wightman's papers.[131] The manuscript focuses on consolidating Wigner's representation theory, rather than introducing new concepts, such as causality. However, others paid close attention both to Wightman's talk and Heisenberg's comment.

5.3 Haag and His Theorem

Wightman made a positive impression during his time in Copenhagen. After meeting him at the Copenhagen conference, Pauli remarked that he had now a "much more positive opinion" of Wightman (Pauli to Møller, August 19, 1952, (von Meyenn, 1996, p. 709)), though he did also make fun of his foundationalist

---

[129] There were also other operationalizations of causality within the framework of the S-Matrix, cf. (Blum and Fraser, 2025).

[130] This connection was soon to be explored in more detail in (Gell-Mann et al., 1954).

[131] I would like to thank Porter Williams for tracking down this manuscript. The manuscript is undated, but it can be dated indirectly through a prominent reference to it: Haag's 1953 lecture notes, mentioned later in this Element, refer to a "survey article by Wigner, Wightman and Bargmann," the content of which appears to match the manuscript in the Wightman papers.

zeal. Pauli picked up on Wightman referring to himself as a missionary (MIW, p. 33), and consequently spelled his name "Whiteman" (von Meyenn, 1996, p. 654). Wightman, the missionary, did actually manage to make one convert, the young German physicist Rudolf Haag. Haag attended the Copenhagen conference as the "delegate" from Munich, which had once been a center of German theoretical physics but had since become somewhat of a provincial backwater. Impressed by Wightman's presentation, Haag recalled:

> Of great importance for my future work was a tea party at the castle where Niels Bohr lived. I got hold of Wightman and we walked many times around the lawn. We saw that we had many aspirations in common and ... he ... recommended in the strongest terms that I should read the 1939 paper by Wigner on the irreducible representations of the inhomogeneous Lorentz group (nowadays called the Poincaré group). After my return from the conference to Munich I dug out this paper [Wigner 1939] and indeed, it was a real revelation. (Haag, 2010b)[132]

Haag returned to Copenhagen in the spring of 1953. In summer, he began working in earnest on relativistic quantum theory. In the academic year 1953/54, he gave a weekly seminar on the subject, resulting in lecture notes, which were published as a preprint of the CERN theory group in Copenhagen in November 1953 (CERN/T/RH-1). Haag characterized his approach as follows:

> We shall not start with a discussion of any particular theory which is at present used in the physics of elementary particles, nor shall we base the investigation upon the general formalism common to all these theories. We begin on a lower level by assuming only some rather general facts of experience and physical principle. The mathematical framework will then be deduced. The advantage of this approach is that we may feel sure of the mathematical consistency ... On the other hand the input is as yet small and consequently the resulting framework wide. However, the comparison with current field theories seems to lead to some interesting statements. (p. 1)

The idea was thus to start from a set of "general principles." The first two of these principles were those of Wigner and Wightman, (A) "the skeleton of quantum mechanics" (states as rays in Hilbert space) and (B') "Invariance of the Laws against Lorentz Transformation." From this starting point, one then gradually added further principles, trying to get as close as possible to regular QFT. It is here that the discussion in this chapter finally makes direct contact with this book's main subject – the consistency of QFT. The final step in Haag's program was to find a model of his axioms, that is, to test their mutual consistency. This model was to consist of a state space, and ultimately of field

---

[132] A similar story is recounted by Wightman, MIW, p. 15.

operators, that obeyed the axioms. Together, field operators and a state space furnish a solution of a QFT, so Haag was ultimately looking for exact solutions, just like Dyson and Källén. But Haag was not trying to get these exact solutions from the dynamical equations; instead, he was trying to short-circuit the dynamics and get the solutions directly from the axioms.

It was clear that axioms (A) and (B′) were not sufficiently constraining. What then were the additional principles that Haag, following Heisenberg's advice, added to those proposed by Wightman? We will get to the principle of causality in a moment. But Haag also introduced another element from the S-Matrix program into the representation-theoretical approach: a focus on the asymptotic behavior of the states.

Recall that Moshinsky had tried to obtain interacting many-particle states by introducing new restrictions on those states in regions where two particles overlap – a natural thing to do when trying to include interactions. Haag now observed that it is far easier to make definite statements about the regions where two (or more) particles should definitely *not* interact anymore, namely the infinite past and future, a common artifice of scattering theory. In these regions, two particles would either form a bound state (and thus be counted as a one-particle state) or ultimately move apart. Haag thus proposed adding an additional simple principle; in Haag's counting, this is Principle 4 because he also included the covariance of states under spatiotemporal reflections as an additional principle alongside Poincaré invariance.[133] Haag's Principle 4 states that, in the asymptotic past and future, the total state of the system will decompose into spatially localized one-particle states, which are localized superpositions of the one-particle momentum eigenstates that Wigner had constructed.

The implications were far-reaching: Haag could now uniquely identify and label each state (no longer just the one-particle states) in the Hilbert space by enumerating the (mass and the spin of the) particles it contained for asymptotic times.[134] In fact, there were *two* possible ways to label the states, corresponding to two possible bases of the Hilbert space, depending on whether one considered the particles in the asymptotic past or the asymptotic future. By identifying

---

[133] Note Haag's unprecedented and unexplainably strange notation: he uses T to denote spatial inversion and C for time reversal. This is particularly odd given that he appears to have been quite attuned to the intricacies of defining time reversal in QFT, such as determining the appropriate formulation of time reversal needed to underwrite the principle of detailed balancing. Around the time that Haag's lecture notes were printed, Gerhart Lüders, another CERN postdoc in Copenhagen, was writing a foundational paper on these matters (see Blum and Martínez de Velasco, 2022).

[134] Adequately symmetrized with either Bose or Fermi statistics (Principle 5) and adding the (unique) no-particle vacuum state in order to complete the Hilbert Space (Principle 6).

not one but two bases for the full Hilbert space of a relativistic quantum theory, Haag effectively completed the Wignerian program. He did this by connecting Wigner's construction of the Hilbert space with the usual state space of QFT, that is, Fock space.

This was a rather natural thing to do and it could be accomplished, as we just saw, by adding a rather uncontroversial principle about asymptotic behavior.[135] However, this success only exacerbated the central problem of the Wigner approach: how to integrate interactions. The scattering amplitudes could now be understood – fully in the spirit of the S-Matrix program – as the unitary transformation that connected the bases corresponding to the asymptotic past and future, respectively. But how were these amplitudes to be determined? Haag had constructed a many-particle Hilbert space without having to invoke a detailed description of the behavior in the region of interaction. Now, some dynamical development was needed to connect the in and out states. Following Heisenberg's anticipation, Haag thus found himself compelled to introduce fields into the picture after all.

Mathematically, this was relatively straightforward. Since he was now in Fock space, Haag could introduce a "complete system" (p. 37) of annihilation and creation operators, $u_\kappa^{(j)}$ and $u_\kappa^{(j)\dagger}$, on his Hilbert space – or rather, two sets, one for the past and one for the future. These operators removed/added asymptotic particles of type (j) (specified by mass and spin) in the state κ (specified by momentum and spin orientation) from/to a given state. In a complete inversion of the usual procedure in field theory, where the field operators were decomposed into annihilation and creation operators, Haag instead used the annihilation and creation operators to construct asymptotic field operators:

$$\varphi^{(j)}(x) = \sum_\kappa f_\kappa^{(j)}(x) u_\kappa^{(j)} + f_\kappa^{*(j)}(x) u_\kappa^{\dagger(j)}, \tag{35}$$

---

[135] Wightman would later claim that he had already discovered in Copenhagen that "in a theory with massive particles, you can write down the unitary equivalence class of the Poincaré group and there is no dynamics in it whatsoever. The same unitary equivalence class if the particles are interacting as when they are not..." (MIW, p. 5–6). This statement suggests that he had obtained results similar to Haag's. There is, however, no documentary evidence for this discovery. The Wigner–Bargmann–Wightman manuscript mentioned earlier does not seem to contain this result, nor does a report by Fritz Coester on a seminar given by Wightman at Princeton provide any such hint: "Wightman gave a couple of seminars on Relativistic Configuration Spaces. In spite of the learned presentation there was no great news in it. He has a program to develop relativistic quantum theory of elementary particles from the particle concepts. As you know I have great sympathy for such an approach. However more than the theory of the Lorentz group is needed to get any non trivial physics and ideas seem to be conspicuously absent" (Letter to Josef Maria Jauch, December 5, 1953, cited in von Meyenn (1999, p. 426)). In any case, Wightman did not contest Haag's priority, remarking that Haag already "knew that intuitively" when they met at the Copenhagen party (MIW, p. 15).

where f is a solution of the free field equations.[136] The S-Matrix could now be interpreted as the transformation connecting the operators of the incoming and outgoing fields. This view was already familiar from conventional perturbation theory; Källén (1950) and Yang and Feldman (1950) had reformulated Dyson's work in the Heisenberg representation and had defined the S-Matrix in that same manner. However, the in and the out fields still lacked a dynamical connection that would allow, at least in principle, the actual calculation of an S-Matrix. Haag proposed to fill this gap with what he called the "interpolating field" $\psi^{(j)}$. This field operator was also constructed from the in- or out-annihilation and creation operators (which, after all, were supposed to form a complete system), but in a more complicated fashion:

$$\psi^{(j)}(x) = \sum_{n,m} \psi^{(j)nm}(x), \qquad (36)$$

where the ψ^{nm} contain normal-ordered products of n creation and m annihilation operators. How exactly this interpolating field was to be calculated was, of course, an open question. Haag did suggest ways to establish at least some contact with regular perturbative calculations (Section IV.3). Nonperturbatively, Haag proposed two conditions to impose on the interpolating field, though he did not elevate these conditions to the level of "principles," perhaps because they would become problematic, as we shall see in a moment. The first condition was called the "asymptotic condition" and was supposed to clarify the relationship between the interpolating field and the asymptotic fields. It stated that the in and out fields emerge as the infinite time limits of the interpolating field, with convergence defined as strong operator convergence,

$$\lim_{t \to \infty} \left\| \left(\psi^{(j)}(x) - \varphi^{(j)}(x) \right) \phi \right\| = 0, \qquad (37)$$

where ϕ is an arbitrary state vector.[137] The second condition was the canonical (equal-time) commutation relations (CCRs) for the interpolating field. These included the microcausality condition discussed earlier, as well as the non-vanishing commutator between the field and its time derivative. While this

---

[136] The reader should note that despite being an asymptotic field, this operator is, in principle, defined for all space-time points.

[137] Haag (2010b) would later point out that this definition of the asymptotic condition was fallacious and that his mistake was corrected by Lehmann, Symanzik and Zimmermann (Lehmann et al., 1955). They imposed an asymptotic condition not on the field operators, but rather on newly defined interpolating creation operators. This point is not relevant to our discussion here. The LSZ amendment is necessary because states such as $\psi^{(j)}(x)\phi$, i.e., states obtained by acting on a normalized state with a field operator, are actually infinite, non-normalizable superpositions, invalidating the convergence condition. This is not the case for the LSZ creation operators. We will discuss this point in somewhat more detail in the context of the work of Léon van Hove.

condition is significantly stronger, all of the subsequent arguments would be equivalent even for mere microcausality (p. 57).

At this point, Haag obtained a result that can be considered the first half of Haag's Theorem. He found that if the interpolating fields were to fulfill the asymptotic condition *and* the CCRs/microcausality, the $\psi^{(j)12}(x)$ in equation 36 would have to vanish. This, in turn, would imply the absence of interactions between (i.e., a trivial S-Matrix for) two incoming particles.[138] What to make of this result? There were three possible conditions for interpolating fields – the asymptotic condition, microcausality, and a nontrivial S-Matrix (i.e., an interacting theory) – and they could not all be simultaneously fulfilled; one of them needed to be dropped. Dropping the demand for an interacting theory – unphysical as that concession was – led to an interesting result: the principles, including Lorentz covariance, the Hilbert space structure of QM and the CCRs/microcausality, were in fact consistent. Taken as axioms, they had a model: the free, noninteracting field operators, which were exact solutions of the free field equations. Haag took this to be the central result of his analysis. He opened the conclusions to his lecture notes:

> It appears that the general formalism of field theory follows essentially from the principles 1–5 formulated in chapters I and II and that a completely consistent mathematical formulation of such a theory is possible... [p. 70]

Proving the in-principle compatibility of special relativity and quantum mechanics was a nontrivial result, even if the fact that the free field equations have exact solutions had, of course, been known from the start. And, in any case, dropping the demand of nontrivial interactions was in no way a viable option. As we have seen, there were two conceivable options that preserved an interacting theory: modifying either the asymptotic condition or the CCRs. Haag seems to have pondered the second option, remarking in his conclusions that "[d]ifficulties arise from the assumption of canonical commutation relations for the field variables." However, it was the first option that gained immediate traction. After all, Haag's formulation of the asymptotic condition was a relatively new invention, whereas microcausality was a hallmark of relativistic QFT.

But if the asymptotic condition was discarded (p. 58), what was left of the connection between interpolating and asymptotic fields? Without such a connection, the introduction of the interpolating fields was pointless. An obvious

---

[138] Of course Haag had not shown that the S-Matrix would be entirely trivial, only the two-particle scattering amplitudes. This was the main desideratum in Haag's proof of his theorem, and later iterations generalized his result to also include more complicated scattering processes (Hall and Wightman, 1957).

idea was that they might be related by some unitary transformation. However, Haag was compelled to reject this option as well. Through very simple arguments (merely invoking translational invariance), he showed that if the asymptotic fields and the interpolating field were connected through a unitary (time evolution) transformation, then that unitary transformation would leave the vacuum invariant. This was, as Haag somewhat cryptically noted, "not fulfilled in the ordinary formulations of electrodynamics or meson theory" (p. 60), probably a reference to the Gell-Mann and Low theorem, which established a nontrivial relationship between the vacuum of the interaction representation (corresponding to the asymptotic fields) and the vacuum of the Heisenberg representation (corresponding to the interpolating fields). Gell-Mann and Low had not investigated whether their transformation was unitary; it certainly involved some nontrivial limits. Could there, then, exist a connection between the asymptotic and the interpolating fields that was neither a limiting relation (like the original asymptotic condition) nor a unitary transformation?

5.4 Friedrichs, Fuglede and Inequivalent Representations

The essential element for the second half of Haag's theorem was the inequivalent representations of the CCRs. In his lecture notes, Haag credited this new mathematical development to two sources: the young Copenhagen mathematician Bent Fuglede and the German-born mathematician K. O. Friedrichs of New York University. Both Friedrichs and Fuglede had recently developed an interest in the mathematical foundations of QFT, likely motivated by the success of renormalized QED in the late 1940s.[139] While Haag appears to have first learned of inequivalent representations from Fuglede, priority for the discovery clearly lies with Friedrichs. So it is with him that we begin.

Friedrichs's main expertise lay in the PDEs of physics (Reid, 1983). Like Wigner and von Neumann, Friedrichs had been shaped by the Göttingen mathematical tradition of Hilbert and Klein. He had been a student of Richard Courant, who had coauthored with Hilbert (and written most of) a famous textbook on *Mathematical Methods of Physics*. In this book, Courant aimed to take "methods – which had their roots in physical intuition," transform them "into rigorous tools backed by general theories" and then "return the improved tools to the physicists for their own work." After losing his Göttingen professorship due to the Nazi racial laws, Courant had left Germany and accepted a position at NYU, where Friedrichs (whose fiancée was Jewish) had joined him in 1937.

---

[139] Fuglede has confirmed to me (phone call, July 19, 2022) that his (ultimately unpublished) results on inequivalent representations were indeed motivated by an interest in quantum field theory. Friedrichs's books is explicitly on the foundations of QFT.

There, they worked together on a second volume of the *Mathematical Methods*, which introduced elements of von Neumann's Hilbert space methods – an approach Courant had not fully absorbed (Reid, 1996, pp. 196–198).

During the war, Friedrichs had worked on the theory of shock waves. Afterward, he became increasingly interested in the mathematics of quantum field theory. In the early 1950s, he engaged with the problem that both Wigner and von Neumann had approached only obliquely, though effectively: providing a mathematical foundation for quantum field theory. Friedrichs was known for a predilection for "the dangerous problems . . ., the really tough ones, in which there was no certainty of getting through" (Reid, 1983). And the foundations of quantum field theory was considered such a problem among mathematicians as evidenced by Irving Segal's review of the book that resulted from Friedrichs's foray into quantum field theory:

> Attempting a mathematical treatment of quantum fields may be a bit like trying to run a cross-country mile in 4 minutes. One of the main obstacles is the psychological one arising from the prevailing opinion that it can't really be done.[140] . . . The subject does not at present admit a systematic or rigorous presentation. To a considerable extent it is not even so much a subject in the mathematical sense as a set of techniques for dealing with specific problems with various elements in common. . . . It is regrettable that the standard of rigor and the mode of presentation are not such as to make the mathematical consistency of many of the sections wholly manifest and explicit, but one must be thankful that the ice has been broken with this first large-scale attempt to deal mathematically with quantum fields. (Segal, 1954)

Friedrichs published his results in a series of long papers in the *Communications on Pure and Applied Mathematics*, which were later combined in a book (Friedrichs, 1953). His approach to quantum theory differed somewhat from the von Neumann tradition, which indirectly, through the Wigner–Wightman connection, also served as the model for Haag's setup. In this latter tradition, the steps in constructing a quantum theory were to start with the Hilbert space, find a complete set of operators, and then deduce the commutation relations. Friedrichs took the opposite approach. He started from the commutation algebra of the observables and then tried to find a representation of this algebra as operators on a Hilbert space. In quantum mechanics, these two approaches were known to be equivalent, due to the Stone–von Neumann theorem; it implied that the canonical commutation relations had a unique (up to unitary transformations) representation as operators on a Hilbert space. One could thus, after one had constructed one complete set of operators from the Hilbert space and

---

[140] Segal's book review was published in November 1954. The mile had been run in less than four minutes for the first time by Roger Bannister in May 1954.

found the commutation relations, take any representation of the commutation relations and again end up with the Hilbert space one had started from. Therefore, it didn't really matter whether one started by setting up the Hilbert space or started with the algebra of the operators. The algebraic approach had already been used before Friedrichs (and even before von Neumann), for example, in the textbook by Weyl (1928, paragraph 46).[141]

Friedrichs now pointed out that things were more complicated in the case of an infinite number of degrees of freedom: here there were also other, inequivalent representations of the commutation relations – be they given as field commutation relations or as commutators of annihilation and creation operators. This was closely connected to von Neumann's results on the nonseparable structure of the state space of QFT: in these inequivalent representations, the annihilation and creation operators acted in the usual way (i.e., lowering and raising occupation numbers, respectively), but on states in different orthogonal subspaces of the larger, nonseparable Hilbert space. In keeping with von Neumann's identification of those other subspaces with states of infinite particle number, Friedrichs distinguished his new representations from the usual one by noting that for the new representations the number operator

$$N = \sum_{K} u_k^{\dagger} u_k, \qquad (38)$$

which corresponds to the total particle number, has an infinite expectation value for any state. Friedrichs consequently referred to the nonstandard representations as "myriotic," because they involved states with a myriad of particles. He also emphasized that such representations were relevant for physics; indeed, it had long been recognized that states with an infinite number of photons with infinitesimally small energies needed to be considered in QED in the context of the infrared divergence.[142]

In the fall of 1953, Fuglede arrived at similar results, in collaboration with – and this may come as a bit of a surprise – Arthur Wightman. Wightman and Fuglede presumably knew each other from Wightman's stay in Copenhagen. Indeed, it appears to have been his interactions with Wightman that briefly sparked Fuglede's interest in QFT. In turn, Wightman was very interested in the question of inequivalent representations, though he did not connect it with the axiomatic approach to QFT; that is why Haag, and not Wightman, is the protagonist of this section. In January 1954, Fuglede decided to abandon this joint project, in order to finish his PhD thesis, giving Wightman

---

[141] For a brief history of the Stone–von Neumann theorem, see (Rosenberg, 2003).
[142] (Bloch and Nordsieck, 1937). On the early history of the infrared divergence, see (Blum, 2015).

permission to complete the paper by himself.[143] Wightman, however, needed a "mathematical technician" (*matematisk tekniker*) to help him with the inequivalent representations, for which he then recruited the Swedish mathematician Lars Gårding, who promptly came to stay at the Institute for Advanced Study from January to April 1954.[144] Wightman wrote two papers with Gårding on inequivalent representations, followed by a third with his former PhD student Sam Schweber (Garding and Wightman, 1954a,b; Wightman and Schweber, 1955).[145] These papers went significantly beyond Friedrichs in their systematic classification of representations of the CCRs. But for us the most important aspect of this story is that, while still actively involved in Wightman's investigation in September 1953, Fuglede also discussed the issue of inequivalent representations with Haag, as he wrote to Wightman:

> [A] few words about the "institute life" here. As usual there are many foreign guests here. One of them, a German whose name is something like Haag, works on problems similar to those you are working with. I have had a few discussions with him concerning some mathematical questions appearing in his investigations. Perhaps you know him?[146]

This is confirmed by Haag, who cites Fuglede as his main source for the existence of inequivalent representations (endnote 18).

We have now established how the inequivalent representations of the CCRs were discovered and how this information reached Haag. But what did it mean for Haag's axiomatic program? For Haag, the existence of the inequivalent representations implied that there were transformations that were not necessarily unitary but still conserved the CCRs. The simplest example, which Haag presented in his lecture notes, is what would later come to be known as a Bogoliubov transformation, where new annihilation and creation operators v and $v^\dagger$

---

[143] Letter from Fuglede to Wightman, January 19, 1954, AWP. I would like to thank Porter Williams for making these letters accessible to me. In this letter, Fuglede also mentions that he was getting married in March, which was another reason for him to stop working on QFT. He also appears to have been somewhat disappointed that his results had already been anticipated by Friedrichs, as he confirmed in a phone conversation on July 19, 2022.

[144] The dates of Gårding's stay can be found on the IAS website (www.ias.edu/scholars/lars-garding). The quote, along with the information that Gårding was invited by Wightman to help with the inequivalent representations, are from Gårding's (Swedish-language) autobiography (Gårding, 2019).

[145] Looking at the chronology of publication, one might think of reading these papers as responses to Haag's theorem, especially since Wightman and Schweber (1955) actually references Haag's lecture notes. In the Arthur Wightman Papers, however, there is a letter to the editor of the *Physical Review*, dated October 14, 1954 (three months after submission), in which Wightman requests "three minor additions to the text as it stands," one of them being the reference to Haag's CERN lectures. Therefore, it seems clear that Haag's lectures could not have influenced Wightman's take on inequivalent representations.

[146] Letter of September 23, 1953, AWP.

are constructed from the annihilation and creation operators u and $u^\dagger$ through the relations:[147]

$$v_\kappa = \cosh \epsilon u_\kappa + \sinh \epsilon u_\kappa^\dagger$$
$$v_\kappa^\dagger = \sinh \epsilon u_\kappa + \cosh \epsilon u_\kappa^\dagger \tag{39}$$

This transformation preserved the CCRs but was not unitary (for an infinite number of states κ), so it delivered a new, not unitarily equivalent representation of the CCRs. The results of Fuglede and Friedrichs therefore supplied Haag with the second part of his theorem: it appeared that the only way to get from his principles to an interacting QFT was to invoke the new inequivalent representations.

The example given by Haag was a rather trivial specimen of a myriotic representation, a "relatively" myriotic representation that actually allows for a Fock space representation with a well-defined number operator, but not within the confines of the original separable Hilbert space.[148] This is not true in general for all myriotic representations. It is unclear whether Haag, at the time, appreciated the distinction; but then again, his theorem[149] also did not indicate what precise kind of nonstandard representations would be needed to construct an interacting QFT. All Haag could conclude was that "the 'strange reps.' ... will almost inevitably play a role in any discussion in field theory" (p. 69).

This posed a challenge for the axiomatic program, which started from the construction of a Hilbert space, only to find that this Hilbert space was too small to accommodate physically interesting theories. The step from establishing the consistency of free QFT as a relativistic quantum theory to the construction of a theory with nontrivial interactions was greater than anticipated. A huge gap remained between the axiomatic exploration of consistency and the study of specific, realistic theories like QED or QFTs of the nuclear interactions. This separation was succinctly summarized by Haag himself several years later at the 1957 conference in Lille on mathematical problems of QFT:

> [A]t the moment there are essentially two distinct methods in use by which one hopes to analyse problems of quantum field theory.

---

[147] Note that in the lecture notes, Haag here had regular sines and cosines rather than hyperbolic ones. This was corrected in the later published version of the notes (Haag, 1955).

[148] For a detailed discussion of these points, and also for a reading of Haag's theorem that fits Haag's original formulation, see (Earman and Fraser, 2006), especially pp. 328ff. Relatively myriotic representations would become essential for the description of spontaneous symmetry breaking. Indeed, Bogoliubov constructed the transformations that now bear his name in the context of superconductivity.

[149] On the naming and subsequent definition of Haag's theorem, see (Lupher, 2005).

> One approach, of which the talk of Källén gives a typical example, starts from an almost completely defined theory (or model). All relevant properties (commutation relations, equations of motion, etc.) are explicitly specified from the outset. However, one is aware of the fact that, in writing down these basic equations, one must make use of mathematical operations which can only be defined a posteriori by a certain limiting procedure. The problem is then to find out just exactly how these mathematical operations should be defined to make the physical predictions of the theory meaningful. Or, if one is less optimistic, to show that no such definition can be given at all.
>
> The other approach has been called 'axiomatic' by Wightman... It was motivated by the following idea. Since there is no inherent contradiction in the notion of a relativistic quantum theory of interacting particles, it seemed worthwhile to start from the wide frame allowed by the general principles and, narrowing the mathematical frame step by step, to look for the neuralgic spots.[150]

The primary difference between the two approaches was thus not so much one of rigor, but of method. For axiomatic QFT, consistency was a principle of construction; for Källén – and the protagonists of the second volume of this Element – it was a tool for analyzing existing structures. A detailed history of axiomatic QFT after the foundational work of Wightman and Haag still remains to be written. While the relative importance that axiomatic field theorists attributed to Haag's theorem fluctuated over time, it remained an essential result for that program. We will conclude our historical treatment of axiomatic QFT here in order to refocus on the consistency of QED. However, we still need to answer one question in this chapter: what was the relevance of Haag's theorem outside of the axiomatic program?

This question was explored by Haag himself, particularly in relation to Dyson's work on the perturbative renormalizability of the S-Matrix. In this work, Dyson had made use of the matrix $U(\sigma)$, a unitary operator that took the wave function at $t = -\infty$ and evolved it to some spacelike hypersurface σ (Dyson, 1949a, pp. 488–489). In the limit where σ is taken to $t = +\infty$, the matrix U is then the S-Matrix and could be made finite through renormalization.

The status of $U(\sigma)$ for finite times was more problematic. Whenever it appeared – whether in the calculation of renormalized Heisenberg field operators for finite times (Dyson, 1951b) or in the context of finite-time evolution (Stueckelberg, 1951)[151] – it had been necessary to blur the initial and final times by averaging over a short time interval. In Haag's framework, the matrix $U(\sigma)$ (without short-time averaging) could be viewed as a unitary transformation

---

[150] (Haag, 1959, p. 151). The original English text was first published as (Haag, 2010a).
[151] On Stueckelberg's results, see also Blum and Fraser (2025).

from the incoming to the interpolating field, a transformation that did not exist for an interacting theory according to Haag's theorem. Haag thus concluded that his results might also explain "the fact that Dyson's U-matrix for finite times is still undefined after renormalization" (p. 67). In the published version of his lecture notes, Haag (1955) used a somewhat less cautious formulation, stating already in the abstract that "Dyson's matrix $U(t_1, t_2)$ for finite t_1 or t_2 cannot exist." However, given his original formulation, one does not get the impression that he felt this entirely invalidated Dyson's derivation – for note that the problem was that U is undefined *after renormalization*. It was therefore impossible to use U to describe finite-time evolution in the renormalized theory. The role of U in the renormalization procedure, however, was not addressed. Haag's attitude seems to have been in keeping with a modern reading where the premises of Haag's theorem are invalidated by the regularizations involved in renormalization, making the use of U as an intermediate step in the derivation of the renormalized S-Matrix far less problematic (Miller, 2015).

In any case, as we have already noted, perturbation theory was essentially dead anyway by the mid 1950s – on a foundational level due to the nonconvergence of the Dyson series, and on a practical level due to its limited usefulness for nuclear interactions. What relevance, then, did Haag's theorem have for the nonperturbative approaches introduced in the previous chapter? Very little, one must say. Haag's theorem was so relevant for the early axiomatic approach because that approach *started* from the construction of the Hilbert space. Even if Haag had come to the conclusion that this representation-theoretical approach could not determine the dynamics, the construction of the Hilbert space remained the first order of business. The nonperturbative approaches we have encountered, on the other hand, started from the dynamical equations. Solutions to these – whether approximate or exact – could be (and in fact were) then sought without first fully defining the state space. As a result, nonseparable Hilbert spaces might arise naturally when solving the dynamical equations. Unbeknownst to Haag, this is essentially what happened a year before Haag's formulation of his theorem, in the work of Léon van Hove.

5.5 Van Hove and the Fixed-Source Model

The Belgian physicist Léon van Hove was yet another mathematically trained newcomer to QFT. In the early 1950s, he began investigating a seemingly minor problem in the foundations of QFT that had recently been pointed out by Hartland Snyder (1950) and Günther Ludwig (1950). They observed that the field operators were not really operators in Hilbert space. When written as an infinite linear combination of annihilation and creation operators, a field operator takes any given state into an infinite, nonnormalizable superposition of states with one particle more than the original state. This is true not just for the field operators but also for interaction Hamiltonians with unpaired field operators, such as the interaction Hamiltonian of QED, which has an unpaired photon operator. Such a Hamiltonian also transforms a given state into an infinite superposition. This difficulty is unrelated to the usual divergence difficulties of QFT; although it can be removed through a cutoff, it cannot be removed through renormalization. While the infinite renormalization of the field operators is proportional to the coupling constant, this problem also arises for free field operators.

As Snyder had pointed out, the problem is ameliorated by the fact that one can formally take the inner product of an infinite superposition with a proper state in Hilbert space, and the result will be finite. In other words, the matrix elements of the field operators will all be finite. The field operators and the Hamiltonian were thus improper operators, akin to the position operator in nonrelativistic quantum mechanics. As long as one is only calculating S-Matrix elements, the problem does not arise, since this only involves calculating matrix elements of the Hamiltonian. If, however, one took a more traditional, quantum-mechanical, approach to QFT and tried to find stationary states – that is, the eigenvalues and eigenvectors of the Hamiltonian – the fact that the Hamiltonian is an improper operator becomes a real problem, because it implies that the Hamiltonian has no eigenvectors in the state space. And while at the avant-garde of the development of QFT the S-Matrix approach ruled supreme, van Hove in 1951 could still confidently assert that:

> The central problem in quantum field theory is to look for the stationary states of the system, i.e., the eigenvectors and eigenvalues of the Hamiltonian operator... (van Hove, 1951, p. 1056)

No one had ever officially renounced this problem. It had become less interesting as scattering experiments displaced spectroscopy. But it was also a prohibitively difficult problem for a relativistic field theory like QED (Blum, 2017). When van Hove's master's (*licence*) student, Marguerite Gossiaux, investigated this problem further in her 1951 thesis, she did so by using a very simplified toy model: a field theory of scalar mesons U interacting with a number of infinitely heavy, point-like nucleons with fixed positions and spin orientations. This was sometimes referred to as the Fixed-Source (or, later, van Hove) model.[152] The interaction Hamiltonian is given by

---

[152] In her thesis, Gossiaux, actually used a somewhat more general model that included a form factor and thus also allowed for extended nucleons. Van Hove later restricted his attention to

$$H_I = g \sum_{s=1}^{l} \gamma_{0,s} U(\vec{x}_s), \qquad (40)$$

where g is the coupling constant, l the total number of nucleons, and $\gamma_{0,s}$ and $\vec{x}_s$ are the Dirac spin matrices and position vectors, respectively, of the nucleons. Both the position and the spin of the nucleons are constants of the motion, and the $\gamma_{0,s}$ can be replaced by their eigenvalues, ± 1, depending on the spin of the nucleon corresponding to the index s.

This Hamiltonian has nontrivial dynamics but is also exactly solvable. Gossiaux was able to formally write down the energy eigenfunctions u of the total Hamiltonian (which also included the kinetic energy of the scalar field) as an infinite product of Hermite polynomials, one for each mode of the scalar field (pp. V.6–V.7). The next question was how the energy eigenfunctions of the interacting theory related to the usual Hilbert space, which is spanned by the energy eigenfunctions of the Hamiltonian for a free, noninteracting meson field. Since the energy eigenfunctions for both the free and interacting theories were now explicitly given, the natural next step was to try to expand one in terms of the other. This meant calculating scalar products. It was at this point that questions of Hilbert space structure first entered – *after* the dynamical equations had been solved.

Solving the Schrödinger equation had required writing the energy eigenfunctions as functions of the variables q_k, the "coordinates" associated with the kth mode of the scalar field. In terms of the usual annihilation and creation operators, a_k and $a_k^\dagger$,

$$q_k = \frac{1}{\sqrt{2}} \left(a_k + a_k^\dagger \right). \qquad (41)$$

This made it natural to perform the scalar product in an infinite product of spaces of square-integrable functions of the variable q_k, which is exactly what Gossiaux tried. However, she had to conclude that

> The expansion coefficients of u in terms of the eigenstates of the free field are infinite products and are given by very complicated expressions.

Gossiaux's discussion of the fixed-source model ended on this inconclusive note. The decisive breakthrough may well have come from Princeton, where van Hove spent the academic year 1949/50 at the Institute for Advanced Study (Messiah, 1993), around the same time Gossiaux was working on her thesis. In

---

point-like sources. This aspect is irrelevant for obtaining exact solutions. However, the van Hove model with extended sources does not exhibit the striking orthogonality properties that van Hove later discovered for the point-source case (Fewster and Rejzner, 2019, Section 3). I would like to thank Maria Papageorgiou for pointing out this latter fact to me.

any case, von Neumann's work on nonseparable Hilbert spaces was still absent from Gossiaux's thesis, while a first citation of von Neumann's 1939 paper appears in the slightly later van Hove (1951).

Van Hove appears to have realized that the state space Gossiaux had implicitly assumed in her thesis was precisely the kind of nonseparable Hilbert space identified by von Neumann.[153] It was therefore much larger than the Hilbert space of the free theory and would include states with infinite particle number. This gave him the decisive pointer: the scalar products that had frustrated Gossiaux could be calculated if, for at least one of the two states, all but a finite number of modes were in the ground state. This greatly simplified the calculation of the scalar products, since the remaining finite number of excited modes would only make a negligible contribution. In this way, van Hove demonstrated that scalar products between energy eigenstates of the interacting and of the free Hamiltonian vanished in general. The eigenstates of the interacting theory thus manifestly lived in a different orthogonal subspace of the full nonseparable space than the eigenvectors of the free Hamiltonian. More than that, if one considered the energy eigenstates for interaction Hamiltonians with different (but nonzero) values of the coupling constant g, these were also elements of mutually orthogonal subspaces.

Going even further, van Hove was then able to show that even infinite-energy eigenstates of the free Hamiltonian, with infinite particle number, were still orthogonal to finite-energy eigenstates of an interacting Hamiltonian. This latter result may be a bit surprising at first glance. It showed that the full nonseparable Hilbert space was not simply obtained by adding states with infinite particle number to the Fock space of the free theory. Van Hove did not elaborate much on this result, but it may be helpful for the reader to understand it better – especially since I was unable to find an elementary explanation in the literature.[154] So I will briefly illustrate van Hove's result in the simplest possible framework, fermionic occupation-number space. Here, the Fock space of the free theory has basis vectors of the form

$$|1\rangle_1 |0\rangle_2 |1\rangle_3 |0\rangle_4 |0\rangle_5 |0\rangle_6 \ldots, \qquad (42)$$

where the indices label the (infinitely many) modes, and the key point is that, beyond some finite index n, all modes are in the ground state $|0\rangle$. To these Fock-space states, one can add further basis vectors with infinite particle number – that is, with an infinite number of modes in the excited state $|1\rangle$. But even with

---

[153] (van Hove, 1952). For an English-language summary of van Hove's paper, see Sbisá (2020).
[154] I would like to thank Maria Papageorgiou, Laura Ruetsche, and, in particular, Noel Swanson for very helpful discussions on this matter.

these infinite-particle states one does not yet have a complete basis for the full nonseparable Hilbert space. One also has to take into account states such as the following:

$$(\alpha|0\rangle_1 + \beta|1\rangle_1)(\alpha|0\rangle_2 + \beta|1\rangle_2)(\alpha|0\rangle_3 + \beta|1\rangle_3)\ldots, \tag{43}$$

where α and β are nonvanishing complex numbers satisfying $|\alpha|^2 + |\beta|^2 = 1$. For each individual mode, such states have a nonzero scalar product with any basis vector of the original Fock space, including basis vectors with infinite particle number. However, that scalar product for each individual mode is always less than one. Thus, the full scalar product, involving an infinite product over all modes, is zero in the end. States such as that of equation (43) thus furnish additional linearly independent basis vectors of the full nonseparable space. And these basis vectors naturally appear as eigenstates of interacting Hamiltonians.

Van Hove had used von Neumann's mathematical insights to resolve the problem of the nature of the field operators and the Hamiltonian – they had to be taken as acting in a much larger space than the Hilbert space of the free theory. In turn, van Hove's result gave physical significance to von Neumann's nonseparable Hilbert space: the orthogonal subspaces were in fact the eigenspaces of Hamiltonians with different interaction terms, where this difference could be as minor as a different numerical value for the coupling constant. Most importantly, the nonseparable Hilbert space arose directly when solving the dynamical equations.

Van Hove's results were soon picked up by the emerging axiomatic QFT community, and cited by Wightman and Schweber (1955) as related to Haag's theorem – albeit not as an actual instance, due to the nonrelativistic nature of the van Hove Model. And outside the axiomatic community, the impact was certainly greater than that of Haag, also because van Hove had explicitly worked out how perturbation theory failed in his model. He performed a perturbative calculation of Dyson's time evolution operator $U(\sigma)$ and obtained nonvanishing amplitudes for the creation of bosons from the vacuum, which he deemed "difficult to interpret." Van Hove's results were sometimes explicitly cited as an objection to perturbation theory, for example, by Alexander Stern (1952), who considered it "more serious" than the nonconvergence of the perturbation series, or by A. John Coleman (1953), who felt that van Hove's result "suggests that... the occasional successes of renormalization methods are lucky flukes." Years later, Källén would still insist to Wightman (letter of January 27, 1964, Källén Papers, University of Lund):

> Preterea censeo: You ought to have mentioned at least van Hove and possibly also Friedrich [sic] in connection with your "Haag" theorem [...] However, I know that you will not change your opinion on that matter.

Given the already problematic state of perturbation theory, we also need to ask what the implications of van Hove's work were for nonperturbative approaches. As I have already indicated, van Hove had shown that nonseparable Hilbert spaces, and the "maze of inequivalent irreducible representations" (Gårding and Wightman, 1954a, p. 617), might well arise naturally when they were needed, thus reducing the relevance of Haag's theorem. But van Hove's contribution was more important on a methodological level: it showed that deep questions about the structure of QFT could be addressed with simplified toy models.[155] In this regard, van Hove's model actually fell slightly short, as it did not involve infinite renormalization (aside from an infinite, constant energy in the Hamiltonian).

This concludes our discussion of the axiomatic tradition and of Haag's theorem – and leads us directly to the theme that will start off the second volume of this Element the emerging methodological tradition of exactly solvable toy models. While Haag was working on his theorem, T. D. Lee was able to construct a model QFT that was superior to van Hove's model in a crucial aspect: it was not just exactly solvable, it also included infinite renormalization. In the second volume, we will discuss how this discovery reinvigorated investigations into the consistency of QED and showed that far worse things could happen with the renormalization constants than simply being infinite.

---

[155] I owe the clear identification of this methodological tradition to Sébastien Rivat (2021), who discusses its relevance to the work of Kenneth Wilson in the 1960s.

Abbreviations

AEA	Albert Einstein Archives, Hebrew University, Jerusalem
AHQP	Archive for the History of Quantum Physics
ASP	Abdus Salam Papers, ICTP Library, Trieste, Italy
ASchP	Alfred Schild Papers, Briscoe Center for American History, Austin, TX
AWP	Arthur Wightman Papers
FRP	Fritz Rohrlich Papers, Niels Bohr Library and Archives in College Park, MD
GKA	Gunnar Källén Archive, Lund University Library, Sweden
JMP	Jagdish Mehra Papers, University of Houston Archives, Houston, TX
JSP	Julian Schwinger Papers, UCLA Library Special Collections, Charles E. Young Research Library Los Angeles, CA
JWP	John Archibald Wheeler Papers, American Philosophical Society Library, Philadelphia, PA
MGP	Murray Gell-Mann Papers, Caltech Archives, Pasadena, CA
NKP	Nicholas Kemmer Papers, Churchill Archives Centre, Cambridge, UK

References

Barrow-Green, J. (2010). The dramatic episode of Sundman. *Historia Mathematica 37*, 164–203.

Belhoste, B. (1991). *Augustin-Louis Cauchy: A Biography*. Berlin: Springer.

Bethe, H. and E. Salpeter (1951). A relativistic equation for bound state problems. *Physical Review 82*, 309–310.

Birkhoff, G. and M. Bennett (1988). Felix Klein and his "Erlanger Programm." In W. Aspray and P. Kitcher (Eds.), *History and Philosophy of Modern Mathematics*, Minnesota Studies in the Philosophy of Science, pp. 145–176. Minneapolis: University of Minnesota Press.

Bloch, F. and A. Nordsieck (1937). Note on the radiation field of the electron. *Physical Review 52*(2), 54–59.

Blum, A. S. (2014). From the necessary to the possible: The genesis of the spin-statistics theorem. *European Physical Journal H 39*, 543–574.

Blum, A. S. (2015). QED and the man who didn't make it: Sidney Dancoff and the infrared divergence. *Studies in History and Philosophy of Modern Physics 50*, 70–94.

Blum, A. S. (2017). The state is not abolished, it withers away: How quantum field theory became a theory of scattering. *Studies in History and Philosophy of Modern Physics 60*, 46–80.

Blum, A. S. (2019). *Heisenberg's 1958 Weltformel and the Roots of Post-Empirical Physics*. Heidelberg: Springer.

Blum, A. S. and J. Fraser (2025). Perturbative causality. *Synthese 206:* 70.

Blum, A. S. and M. Jähnert (2024). Quantum mechanics, radiation, and the equivalence proof. *Archive for History of Exact Sciences 78*(5), 567–616.

Blum, A. S. and A. Martínez de Velasco (2022). The Genesis of the CPT Theorem. *The European Physical Journal H 47*, article 5.

Blum, A. S. and D. Rickles (eds.) (2018). *Quantum Gravity in the First Half of the XXth Century: A Sourcebook*. Berlin: Edition Open Access.

Bogoliubov, N. and D. Shirkov (1959). *Introduction to the Theory of Quantized Fields*. London: Interscience.

Bohr, N. (1932). Chemistry and the quantum theory of atomic constitution. *Journal of the Chemical Society of London*, 349–384.

Bohr, N. (1936). Neutron capture and nuclear constitution. *Nature 137*, 344–368.

Borowitz, S. and W. Kohn (1949). On the electromagnetic properties of nucleons. *Physical Review 76*(6), 818–827.

Browder, F. (ed.) (1976). *Mathematical Developments Arising from Hilbert Problems*. Providence, RI: American Mathematical Society.

Cao, T. Y. (2010). *From Current Algebra to Quantum Chromodynamics*. Cambridge: Cambridge University Press.

Cao, T. Y. and S. S. Schweber (1993). The conceptual foundations and the philosophical aspects of renormalization theory. *Synthese 97*, 33–108.

Carrington, M. E., H. Defu, and M. H. Thoma (1999). Equilibrium and non-equilibrium hard thermal loop resummation in the real time formalism. *The European Physical Journal C 7*, 347–354.

Cauchy, A. (1840). Mémoire sur l'intégration des équation différentielles. In *Exercices d'Analyse et de Physique Mathématique. Tome Premier.*, pp. 327–384. Paris: Bachelier.

Close, F. (2011). *The Infinity Puzzle*. Oxford: Oxford University Press.

Coleman, A. J. (1953). Review of L. van Hove's "Les difficultés de divergences pour un modèle particulier de champ quantifié." *Mathematical Reviews*.

Cooke, R. (1984). *The Mathematics of Sonya Kovalevskaya*. New York: Springer.

Corry, L. (2004). *David Hilbert and the Axiomatization of Physics*. Dordrecht: Kluwer.

Cushing, J. T. (1990). *Theory Construction and Selection in Modern Physics: The S Matrix*. Cambridge: Cambridge University Press.

Darrigol, O. (2005). *Worlds of Flow*. Oxford: Oxford University Press.

Dirac, P. (1927a). The quantum theory of dispersion. *Proceedings of the Royal Society of London A114* (769), 710–728.

Dirac, P. A. M. (1927b). The quantum theory of emission and absorption of radiation. *Proceedings of the Royal Society of London A114*, 243–265.

Dirac, P. A. M. (1928a). The quantum theory of the electron. *Proceedings of the Royal Society of London A117*: 610–624.

Dirac, P. A. M. (1928b). The quantum theory of the electron. Part II. *Proceedings of the Royal Society of London A 118*, 351–361.

Dirac, P. A. M. (1930). *The Principles of Quantum Mechanics*. Oxford: Clarendon Press.

Dirac, P. (1987). The inadequacies of quantum field theory. In B. Kursunoglu and E. Wigner (Eds.), *The Dirac Memorial Volume*, pp. 194–198. Cambridge: Cambridge University Press.

Duncan, A. (2012). *The Conceptual Framework of Quantum Field Theory*. Oxford: Oxford University Press.

Dunne, G. V. and M. Ünsal (2012). Resurgence and trans-series in quantum field theory: The CP^{N-1} model. *Journal of High Energy Physics article 170*.

Dyson, F. (1952). Divergence of perturbation theory in quantum electrodynamics. *Physical Review 85*(4), 631–632.

Dyson, F. (1979). *Disturbing the Universe*. New York: Harper & Row.

Dyson, F. (2018). *Maker of Patterns: An Autobiography through Letters*. New York: Liveright.

Dyson, F. J. (1949a). The radiation theories of Tomonaga, Schwinger, and Feynman. *Physical Review 75*(3), 486–502.

Dyson, F. J. (1949b). The S matrix in quantum electrodynamics. *Physical Review 75*(11), 1736–1755.

Dyson, F. J. (1951a). Heisenberg operators in quantum electrodynamics. I. *Physical Review 82*(3), 428–439.

Dyson, F. J. (1951b). The renormalization method in quantum electrodynamics. *Proceedings of the Royal Society of London A 207*(1090), 395–401.

Dyson, F., M. Ross, E. Salpeter, S. Schweber, M. Sundaresan, W. Visscher, and H. Bethe (1954). Meson-nucleon scattering in the Tamm–Dancoff Approximation. *Physical Review 95*(6), 1644–1658.

Earman, J. and D. Fraser (2006). Haag's theorem and its implications for the foundations of quantum field theory. *Erkenntnis 64*(3), 305–344.

Eden, R. J. (1952). Threshold behaviour in quantum field theory. *Proceedings of the Royal Society of London A 210*(1102), 388–404.

Edwards, S. (1953). A non-perturbation approach to quantum electrodynamics. *Physical Review 90*(2), 284–291.

Einstein, A. (1905). Über einen die Erzeugung und die Verwandlung des Lichtes betreffenden heuristischen, Gesichtspunkt. *Annalen der Physik 17*, 132–148.

Fermi, E. (1934). Versuch einer Theorie der β-Strahlen. I. *Zeitschrift für Physik 88*(3), 161–177.

Fewster, C. J. and K. Rejzner (2019, November). Algebraic quantum field theory: An introduction. arXiv:1904.04041v2 [hep-th].

Feynman, R. P. (1949). The theory of positrons. *Physical Review 76*(6), 749–759.

Feynman, R. P. (1985). *QED: The Strange Theory of Light and Matter*. Princeton, NJ: Princeton University Press.

Fock, V. (1932). Konfigurationsraum und zweite Quantelung. *Zeitschrift für Physik 75*, 622–647.

Fraser, D. (2011). How to take particle physics seriously: A further defence of axiomatic quantum field theory. *Studies in History and Philosophy of Modern Physics 42*(2), 126–135.

Fraser, G. (2008). *Cosmic Anger: Abdus Salam – The First Muslim Nobel Scientist*. Oxford: Oxford University Press.

Fraser, J. D. (2016). What is Quantum Field Theory? Idealisation, Explanation and Realism in High Energy Physics. PhD thesis, University of Leeds.

Fraser, J. D. (2021). The twin origins of renormalization group concepts. *Studies in History and Philosophy of Science 89*, 114–128.

Fried, H. M. (1959). Structure theorem for the photon propagator. *Physical Review 115*(1), 220–222.

Friedrichs, K. (1953). *Mathematical Aspects of the Quantum Theory of Fields*. New York: Interscience.

Frisch, M. (2005). *Inconsistency, Asymmetry, and Non-Locality: A Philosophical Investigation of Classical Electrodynamics*. Oxford: Oxford University Press.

Froissart, M. (1961). Asymptotic behavior and subtractions in the Mandelstam representation. *Physical Review 123*(3), 1053–1057.

Gårding, L. (2019). *Levnadsbeskrivning*.

Gårding, L. and A. S. Wightman (1954a). Representations of the anticommutation relations. *Proceedings of the National Academy of Sciences 40*(7), 617–621.

Gårding, L. and A. S. Wightman (1954b). Representations of the commutation relations. *Proceedings of the National Academy of Sciences 40*(7), 622–626.

Gell-Mann, M. (1956). The interpretation of the new particles as displaced charge multiplets. *Supplemento al Nuovo Cimento 4*(2), 848–866.

Gell-Mann, M., M. Goldberger, and W. Thirring (1954). Use of causality conditions in quantum theory. *Physical Review 95*, 1612–1627.

Gell-Mann, M. and F. Low (1951). Bound states in quantum field theory. *Physical Review 84*(2), 350–354.

Gell-Mann, M. and F. Low (1954). Quantum electrodynamics at small distances. *Physical Review 95*(5), 1300–1312.

Gross, D. (1990). Chasing the Landau Ghost. In E. Gotsman, Y. Ne'eman, and A. Voronel (Eds.), *Frontiers of Physics: Proceedings of the Landau Memorial Conference, Tel Aviv, Israel, 6–10 June 1988*, pp. 97–111. Oxford: Pergamon Press.

Haag, R. (1955). On quantum field theories. *Det Kgl. Danske Videnskabernes Selskab Mathematisk-Fysiske Meddelelser 29*(12), 1–37.

Haag, R. (1959). Discussion des "axiomes" et des propriétés asymptotique d'une théorie des champs locale avec particules composées. In *Les Problèmes Mathématiques de la Théorie Quantique des Champs*, pp. 151–162. Paris: Centre National de la Recherche Scientifique.

Haag, R. (2010a). Discussion of the 'axioms' and the asymptotic properties of a local field theory with composite particles. *European Physical Journal H 35*(3), 243–253.

Haag, R. (2010b). Some people and some problems met in half a century of commitment to mathematical physics. *European Physical Journal H 35*(3), 263–307.

Hadamard, J. (1907). Les problèmes aux limites dans la théorie des équations aux dérivées partielles. *Journal de Physique Théorique et Appliquée 6*(1), 202–241.

Hadamard, J. (1923). *Lectures on Cauchy's Problem in Linear Partial Differential Equations*. New Haven, CT: Yale University Press.

Hall, D. and A. Wightman (1957). A theorem on invariant analytic functions with applications to relativistic quantum field theory. *Danske Videnskabernes Selskab Matematisk-Fysiske Meddelelser 31*(5), 1–41.

Hamilton, A. (2009). James Hamilton, Physicist: Biography of Irish Physicist James Hamilton. jameshamiltonphysicist.com.

Hamilton, J. (1986). On mesons and methods. *Arkhimedes 38*, 118–141.

Hanneke, D., S. Fogwell Hoogerheide, and G. Gabrielse (2011). Cavity control of a single-electron quantum cyclotron: Measuring the electron magnetic moment. *Physical Review A 83*, 052122–1–26.

Hardy, G. (1949). *Divergent Series*. Oxford: Clarendon Press.

Heisenberg, W. (1931). Bemerkungen zur Strahlungstheorie. *Annalen der Physik 401*(3), 338–346.

Heisenberg, W. and W. Pauli (1929). Zur Quantendynamik der Wellenfelder. *Zeitschrift für Physik 56*, 1–61.

Heitler, W. (1936). *The Quantum Theory of Radiation* (1st ed.). Oxford: Clarendon Press.

Hermann, A., K. von Meyenn, and V. F. Weisskopf (eds.) (1979). *Wolfgang Pauli: Wissenschaftlicher Briefwechsel mit Bohr, Einstein, Heisenberg u.a.*, Volume I: 1919–1929. New York: Springer.

Hertz, H. (1893). *Electric Waves*. London: Macmillan.

Hilbert, D. (1915). Die Grundlagen der Physik. Erste Mitteilung. *Nachrichten von der Gesellschaft der Wissenschaften zu Göttingen, Mathematisch-Physikalische Klasse*, 395–408.

Hurst, C. (1952a). An example of a divergent perturbation expansion in field theory. *Mathematical Proceedings of the Cambridge Philosophical Society* 48(4), 625–639.

Hurst, C. (1952b). The enumeration of graphs in the Feynman–Dyson technique. *Proceedings of the Royal Society of London A 214*(1116), 44–61.

Jarlskog, C. (2014). *Portrait of Gunnar Källén*. Cham: Springer.

Jauch, J. M. and F. Rohrlich (1955). *The Theory of Photons and Electrons*. Reading, MA: Addison-Wesley.

Jordan, P. (1933). Über Verallgemeinerungsmöglichkeiten des Formalismus der Quantenmechanik. *Nachrichten der Gesellschaft der Wissenschaften zu Göttingen*, 209–217.

Jordan, P. and J. von Neumann (1935). On inner products in linear, metric spaces. *Annals of Mathematics*, 719–723.

Kaiser, D. (2005). *Drawing Theories Apart: The Dispersion of Feynman Diagrams in Postwar Physics*. Chicago, IL: University of Chicago Press.

Källén, G. (1950). Mass and charge-renormalizations in quantum electrodynamics without the use of the interaction representation. *Arkiv för Fysik 2*, 187–194.

Källén, G. (1952). On the definition of the renormalization constants in quantum electrodynamics. *Helvetica Physica Acta 25*, 417–434.

Källén, G. (1953). On the magnitude of the renormalization constants in quantum electrodynamics. *Det Kgl. Danske Videnskabernes Selskab Mathematisk-fysiske Meddelelser 27*(12), 1–18.

Källén, G. (1955). Review of Gell-Mann, M. and F. E. Low: Quantum electrodynamics at small distances. *Zentralblatt für Mathematik 57*, 214.

Kamefuchi, S. (1951). Note on the direct interaction between spinor fields. *Progress of Theoretical Physics 6*(2), 175–181.

Karplus, R. and A. Klein (1952). Electrodynamic displacement of atomic energy levels. III. The hyperfine structure of positronium. *Physical Review 87*(5), 848–858.

Klein, F. (1882). *Über Riemann's Theorie der Algebraischen Functionen und ihrer Integrale*. Leipzig: Teubner.

Klein, F. (1926). *Vorlesungen über die Entwicklung der Mathematik im 19. Jahrhundert*. Berlin: Springer.

Klein, O. and Y. Nishina (1929). Über die Streuung von Strahlung durch freie Elektronen nach der neuen relativistischen Quantendynamik von Dirac. *Zeitschrift für Physik 52*(11-12), 853–868.

Kline, M. (1972). *Mathematical Thought from Ancient to Modern Times*. Oxford: Oxford University Press.

Koberinski, A. (2021). Mathematical development in the rise of Yang–Mills gauge theories. *Synthese 198*, S3747–S3777.

Kofoed-Hansen, O., P. Kristensen, M. Scharff, and A. Winther (Eds.) (1952). *Report of the International Physics Conference – Institute for Theoretical Physics, Copenhagen, June 3-17, 1952*.

Kragh, H. (2003). Magic number: A partial history of the fine-structure constant. *Archive for History of Exact Sciences 57*, 395–431.

Kronig, R. (1946). A supplementary condition in Heisenberg's theory of elementary particles. *Physica 12*, 543–544.

Landau, L. and R. Peierls (1930). Quantenelektrodynamik im Konfigurationsraum. *Zeitschrift für Physik 62*, 188–200.

Laplace, P. (1798). *Traité de Mécanique Céleste*. Paris: J. B. M. Duprat.

Lee, S. (Ed.) (2009). *Sir Rudolf Peierls: selected private and scientific correspondence*, Volume 2. Hackensack, NJ: World Scientific.

Lehmann, H. (1954). Über Eigenschaften von Ausbreitungsfunktionen und Renormierungskonstanten quantisierter Felder. *Il Nuovo Cimento 11*, 342–357.

Lehmann, H., K. Symanzik, and W. Zimmermann (1955). Zur Formulierung quantisierter Feldtheorien. *Il Nuovo Cimento 1*(1), 205–225.

Li, B. (2015). Coarse-graining as a route to microscopic physics: The renormalization group in quantum field theory. *Philosophy of Science 82*, 1211–1223.

Ludwig, G. (1950). Ansatz zu einer divergenzfreien Quantenelektrodynamik. *Zeitschrift für Naturforschung A 5a*(12), 637–641.

Lupher, T. (2005). Who Proved Haag's Theorem? *International Journal of Theoretical Physics 44*(11), 1995–2005.

Majer, U. (2001). The axiomatic method and the foundations of science: Historical Roots of Mathematical Physics in Göttingen (1900–1933). In M. Rédei and M. Stoelzner (Eds.), *John von Neumann and the Foundations of Quantum Physics*, pp. 11–33. Dordrecht: Springer.

Martin, P. C. (1979). Schwinger and statistical physics: a spin-off success story and some challenging sequels. *Physica 96A*, 70–88.

Matsubara, Y., T. Suzuki, and I. Yotsuyanagi (1984). A Constraint from the Källén–Lehmann Representation on the $(\phi^4)_4$ Theory. *Progress of Theoretical Physics 72*(4), 873–876.

Matthews, P. (1949). The application of Dyson's method to meson interactions. *Physical Review 76*, 684–685.

Matthews, P. (1950). The S-matrix for meson-nucleon interactions. *Philosophical Magazine 41*(313), 185–195.

Matthews, P. and A. Salam (1951). The renormalization of meson theories. *Reviews of Modern Physics 23*(4), 311–314.

Matthews, P. and A. Salam (1954). Renormalization. *Physical Review 94*(1), 185–191.

Meheus, J. (Ed.) (2002). *Inconsistency in Science*. Dordrecht: Kluwer.

Messiah, A. (1993). The early post-War period of Léon van Hove. A recollection. In F. Nicodemi (Ed.), *Scientific Highlights in Memory of Léon van Hove*. Singapore: World Scientific.

Miller, M. E. (2015). Haag's theorem, apparent inconsistency, and the empirical adequacy of quantum field theory. *British Journal for the Philosophy of Science*.

Miller, M. E. (2017). The Structure and Interpretation of Quantum Field Theory. PhD thesis, University of Pittsburgh. https://d-scholarship.pitt.edu/328 18/1/miller_etd_2017_2.pdf.

Monna, A. (1975). *Dirichlet's Principle: A Mathematical Comedy of Errors and Its Influence on the Development of Analysis*. Utrecht: Oosthowk, Scheltema & Holkema.

Moshinsky, M. (1949, January). *Relativistic Interactions by Means of Boundary Conditions*. PhD thesis, Princeton University.

Nambu, Y. (1955). Structure of Green's Functions in Quantum Field Theory. *Physical Review 100*(1), 394–411.

Nickles, T. (2002). From Copernicus to Ptolemy: Inconsistency and method. In J. Meheus (Ed.), *Inconsistency in Science*, pp. 1–33. Dordrecht: Kluwer.

Oppenheimer, J. R. (1930). Note on the theory of the interaction of field and matter. *Physical Review 35*(5), 461–477.

Oppenheimer, J. R. (1950). Electron theory. In *Les Particules Élémentaires. Proceedings of the 8th Solvay Conference on Physics*, pp. 269–286. Brussels: R. Stoops.

Pauli, W. and F. Villars (1949). On the invariant regularization in relativistic quantum theory. *Reviews of Modern Physics 21*, 434–444.

Petermann, A. (1953). Divergence of perturbation expansion. *Physical Review 89*(5), 1160–1161.

Poincaré, H. (1993). *New Methods in Celestial Mechanics*. College Park, MD: American Institute of Physics.

Rédei, M. and M. Stöltzner (Eds.) (2001). *John von Neumann and the Foundations of Quantum Physics*. Berlin: Springer.

Rédei, M. and M. Stöltzner (2006). Soft axiomatisation: John von Neumann on method and von Neumann's method in the physical sciences. In E. Carson and R. Huber (Eds.), *Intuition and the Axiomatic Method*, pp. 235–249. Dordrecht: Springer.

Reid, C. (1983). K. O. Friedrichs 1901–1982. *The Mathematical Intelligencer 5*, 23–30.

Reid, C. (1996). *Courant*. New York: Copernicus.

Rickles, D. (2020). *Covered with Deep Mist: The Development of Quantum Gravity (1916–1956)*. Oxford: Oxford University Press.

Rivat, S. (2021). Drawing scales apart: The origins of Wilson's conception of effective field theories. *Studies in History and Philosophy of Science 90*, 321–338.

Rosenberg, J. (2003). A selective history of the Stone–von Neumann theorem. In R. S. Doran and R. V. Kadison (Eds.), *Operator Algebras, Quantization, and Noncommutative Geometry: A Centennial Celebration Honoring John von Neumann and Marshall H. Stone*. Providence, RI: American Mathematical Society.

Rosenfeld, L. (1930). Über die Gravitationswirkungen des Lichtes. *Zeitschrift für Physik 65*, 589–599.

Rosenfeld, L. (1931). Zur Kritik der Diracschen Strahlungstheorie. *Zeitschrift für Physik 70*, 454–462.

Rosenfeld, L. (1932). Über die quantentheoretische Behandlung der Strahlungsprobleme. In *Convegno di Fisica Nucleare, Ottobre 1931 - IX, Roma*, Volume 1 of *Atti di Convegni*, pp. 131–135. Reale Academia d'Italia.

Rueger, A. (1992). Attitudes towards infinities: Responses to anomalies in quantum electrodynamics, 1927–1947. *Historical Studies in the Physical and Biological Sciences 22*, 309–337.

Sakata, S. (1947). Correlation between mutual interactions of elementary particles. *Progress of Theoretical Physics 2*(2), 99.

Salpeter, E. and H. Bethe (1951). A relativistic equation for bound-state problems. *Physical Reviewl 84*, 1232–1242.

Salpeter, E. E. (2008). Bethe–Salpeter equation (origins). *Scholarpedia 3*(11), 7483.

Sbisà, F. (2020). On Léon van Hove's 1952 article on the foundations of Quantum Field Theory. arXiv: 2010.02199v1.

Schlissel, A. (1976). The development of asymptotic solutions of linear differential equations. *Archive for History of Exact Sciences 16*, 307–378.

Schweber, S. S. (1994). *QED and the Men Who Made It: Dyson, Feynman, Schwinger, and Tomonaga*. Princeton, NJ: Princeton University Press.

Schweber, S. S., H. A. Bethe, and F. de Hoffmann (1955). *Mesons and Fields*. Evanston, IL: Row, Peterson and Company.

Schwinger, J. (1948). On quantum electrodynamics and the magnetic moment of the electron. *Physical Review 73*, 416–417.

Schwinger, J. (1951a). On the Green's functions of Quantized Fields. I. *Proceedings of the National Academy of Sciences 37*(7), 452–455.

Schwinger, J. (1951b). On the Green's functions of Quantized Fields. II. *Proceedings of the National Academy of Sciences 37*(7), 455–459.

Schwinger, J. (1951c). The theory of quantized fields, I. *Physical Review 82*, 914–927.

Schwinger, J. (1996). The Greening of quantum field theory: George and I. In Y. J. Ng (Ed.), *Julian Schwinger: The Physicist, the Teacher, and the Man*, pp. 13–27. Singapore: World Scientific.

Segal, I. (1954). Review of "Mathematical aspects of the quantum theory of fields" (by K. O. Friedrichs). *Bulletin of the American Mathematical Society 60*, 569–570.

Snyder, H. (1950). Quantum field theory. *Physical Review 79*(3), 520–525.

Stern, A. W. (1952). Space, field, and ether in contemporary physics. *Science*, 493–496.

Stueckelberg, E. C. G. (1951). Relativistic quantum theory for finite time intervals. *Physical Review 81*(1), 130–133.

Stueckelberg, E. C. G. and A. Petermann (1953). La normalisation des constantes dans la théorie des quanta. *Helvetica Physica Acta 26*(5), 499–520.

Symanzik, K. (1954). Über das Schwingersche Funktional in der Feldtheorie. *Zeitschrift für Naturforschung 9a*, 809–824.

't Hooft, G. (1976). Symmetry Breaking through Bell–Jackiw Anomalies. *Physical Review Letters 37*(1), 8–11.

Thirring, W. (1953). On the divergence of perturbation theory for quantized fields. *Helvetica Physica Acta 26*, 33–52.

Thirring, W. (2008). *Lust am Forschen*. Vienna: Seifert Verlag.

Thirring, W. E. (1951). Zum Wert der Renormalisationskonstanten. *Zeitschrift für Naturforschung 6a*, 462–463.

Truesdell, C. A. (Ed.) (1954). *Leonhardi Euleri Opera Omnia II: 12*, Volume II. Zurich: Orell Füssli.

Uehling, E. A. (1935). Polarization effects in the positron theory. *Physical Review 28*(1), 55–63.

Umezawa, H. and S. Kamefuchi (1951). The vacuum in quantum electrodynamics. *Progress of Theoretical Physics 6*(4), 543–558.

Utiyama, R., S. Sunakawa, and T. Imamura (1952). On the theory of the Green-functions in quantum electrodynamics. *Progress of Theoretical Physics 8*(1), 77–110.

van Hove, L. (1951). Sur l'opérateur Hamiltonien de deux champs quantifiés en interaction. *Bulletin de la Classe des Sciences 37*, 1055–1072.

van Hove, L. (1952). Les difficultés de divergences pour un modèle particulier de champ quantifié. *Physica 18*(3), 10–24.

Vickers, P. (2013). *Understanding inconsistent science*. Oxford: Oxford University Press.

von Helmholtz, H. (1873). Ueber ein Theorem, geometrisch ähnliche Bewegungen flüssiger Körper betreffend, nebst Anwendung auf das Problem, Luftballons zu lenken. *Monatsberichte der Königlichen Akademie der Wissenschaften zu Berlin*, 501–514.

von Meyenn, K. (Ed.) (1993). *Wolfgang Pauli: Wissenschaftlicher Briefwechsel mit Bohr, Einstein, Heisenberg u.a.*, Volume III: 1940–1949. Berlin: Springer.

von Meyenn, K. (Ed.) (1996). *Wolfgang Pauli: Wissenschaftlicher Briefwechsel mit Bohr, Einstein, Heisenberg u.a.*, Volume IV/Part I: 1950–1952. Berlin: Springer.

von Meyenn, K. (Ed.) (1999). *Wolfgang Pauli: Wissenschaftlicher Briefwechsel mit Bohr, Einstein, Heisenberg u.a.*, Volume IV/Part II: 1953–1954. Berlin: Springer.

von Meyenn, K. (Ed.) (2001). *Wolfgang Pauli: Wissenschaftlicher Briefwechsel mit Bohr, Einstein, Heisenberg u.a.*, Volume IV/Part III: 1955–1956. Berlin: Springer.

von Neumann, J. (1929). Allgemeine Eigenwerttheorie Hermitescher Funktionaloperatoren. *Mathematische Annalen 102*, 49–131.

von Neumann, J. (1932). *Mathematische Grundlagen der Quantenmechanik*. Berlin: Springer.

von Neumann, J. (1939). On infinite direct products. *Compositio Mathematica 6*, 1–77.

von Neumann, J. (1951). *The Geometry of Orthogonal Spaces*. Princeton, NJ: Princeton University Press.

von Weizsäcker, C. F. (1999). *Große Physiker*. Hanser.

Wallace, D. (2011). Taking particle physics seriously: A critique of the algebraic approach to quantum field theory. *Studies in History and Philosophy of Modern Physics 42*(2), 116–125.

Waller, I. (1930). Bemerkungen über die Rolle der Eigenenergie des Elektrons in der Quantentheorie der Strahlung. *Zeitschrift für Physik 62*(9-10), 673–676.

Ward, J. (1951). Renormalization theory of the interactions of nucleons, mesons, and photons. *Physical Review 84*, 897–901.

Ward, J. (2004). *Memoirs of a Theoretical Physicist*. Rochester, NY: Optics Journal.

Ward, J. C. (1950). A convergent non-linear field theory. *Physical Review 79*(2), 406.

Weinberg, S. (1995). *The Quantum Theory of Fields*, Volume I: Foundations. Cambridge: Cambridge University Press.

Weyl, H. (1928). *Gruppentheorie und Quantenmechanik* (1st ed.). Leipzig: Hirzel.

Wick, G. (1950). The evaluation of the collision matrix. *Physical Review 80*(2), 268–272.

Wightman, A. (1952). An example of a consistent relativistic quantum theory of interacting particles. In *Report of the International Physics Conference – Institute for Theoretical Physics, Copenhagen, June 3–17, 1952*.

Wightman, A. (1976). Hilbert's sixth problem: Mathematical treatment of the axioms of physics. In F. Browder (ed.), *Mathematical Developments arising from Hilbert Problems*, Proceedings of Symposia in Pure Mathematics, pp. 148–240. Providence, RI: American Mathematical Society.

Wightman, A. (1979). Should we believe in quantum field theory? In A. Zichichi (Ed.), *The Whys of Subnuclear Physics*, pp. 983–1025. New York: Plenum Press.

Wightman, A. S. (1956). Quantum field theory in terms of vacuum expectation values. *Physical Review 101*(2), 860–866.

Wightman, A. S. and S. S. Schweber (1955). Configuration space methods in relativistic quantum field theory. I. *Physical Review 98*(3), 812–837.

Wigner, E. (1939). On unitary representations of the inhomogeneous Lorentz group. *Annals of Mathematics 40*(1), 149–204.

Wigner, E. (1948). Relativistische Wellengleichungen. *Zeitschrift für Physik 124*(7), 665–684.

Wilson, K. G. (1971). Renormalization group and strong interactions. *Physical Review D 3*(8), 1818–1846.

Yang, C.-N. and D. Feldman (1950). The S-matrix in the Heisenberg representation. *Physical Review 79*(6), 972–978.

Yukawa, H. (1935). On the interaction of elementary particles. I. *Proceedings of the Physico-Mathematical Society of Japan 17*, 48–57.

Cambridge Elements =

Foundations of Contemporary Physics

Richard Dawid
Stockholm University

Richard Dawid is Professor in the Philosophy of Science at Stockholm University and specialises in the philosophy of contemporary physics, particularly that of non-empirical theory assessment. In 2013 he published *String Theory and the Scientific Method* and in 2019 he co-edited a second book titled *Why Trust a Theory* (both published by Cambridge University Press).

James Wells
University of Michigan, Ann Arbor

James Wells is Professor in Physics at the University of Michigan, Ann Arbor, and his research specialises in theoretical cosmology, with a particular focus on foundational questions in fundamental physics such as gauge symmetries, CP violation, naturalness and cosmological history. He is a fellow of the American Physical Society.

About the Series

Foundations in Contemporary Physics explores some of the most significant questions and discussions currently taking place in modern physics. The series is accessible to physicists and philosophers and historians of science, and has a strong focus on cutting-edge topics of research such as quantum information, cosmology, and big data.

Cambridge Elements

Foundations of Contemporary Physics

Elements in the Series

Quantum Gravity in a Laboratory?
Nick Huggett, Niels Linnemann, and Mike D. Schneider

Gauge Symmetries, Symmetry Breaking, and Gauge-Invariant Approaches
Philipp Berghofer, Jordan François, Simon Friederich, Henrique Gomes, Guy Hetzroni, Axel Maas, and René Sondenheimer

Probing the Consistency of Quantum Field Theory I: From Nonconvergence to Haag's Theorem (1949–1954)
Alexander S. Blum

A full series listing is available at: www.cambridge.org/EFCP

For EU product safety concerns, contact us at Calle de José Abascal, 56–1°,
28003 Madrid, Spain or eugpsr@cambridge.org.

www.ingramcontent.com/pod-product-compliance
Lightning Source LLC
LaVergne TN
LVHW011726060526
838200LV00051B/3048